TECHNOLOGY, INNOVATION and POLICY 9

Series of the Fraunhofer Institute
for Systems and Innovation Research (ISI)

TECHNOLOGICAL INNOVATION and POLICY 2

Frieder Meyer-Krahmer (Ed.)

Globalisation of R&D and Technology Markets

Consequences for National Innovation Policies

Proceedings of the International Conference
on 1 and 2 December 1997 in Bonn, Petersberg, Germany
organised by the BMBF and the FhG ISI

**With 51 Figures
and 11 Tables**

Physica-Verlag

A Springer-Verlag Company

Professor Dr. Frieder Meyer-Krahmer
Fraunhofer Institute for
Systems and Innovation Research (ISI)
Breslauer Str. 48
D-76139 Karlsruhe, Germany
and
Université Louis Pasteur
Strasbourg, France

ISBN 978-3-7908-1175-9   ISBN 978-3-642-49957-9 (eBook)
DOI 10.1007/978-3-642-49957-9
Die Deutsche Bibliothek – CIP-Einheitsaufnahme
Globalisation of R & D and technology markets : consequences
for national innovation policies; with 11 tables; proceedings of the
international conference on 1 and 2 December 1997 in Bonn,
Petersberg, Germany / organised by BMBF and the FhG ISI. Frieder
Meyer-Krahmer (ed.). - Heidelberg; New York : Physica-Verl., 1999
(Technology, innovation and policy; Vol. 9)

Cover design: Erich Kirchner, Heidelberg
SPIN 10697671   88/2202-5 4 3 2 1 0-Printed on acid-free paper

Table of Contents

3. The Firm's Perspective

1. Introduction and Policy Perspective

Introduction

Frieder Meyer-Krahmer

On the threshold of the 21st century an intensive debate about globalisation of R&D and technology markets is of crucial importance. Globalisation will carry on changing the framework for company strategies and subsequently for public policies. It will, in fact, contribute to global prosperity. However, the winners in a closely interlinked world economy will probably be those locations which, owing to competence and openness, become centres of information, communication and knowledge application. It is the overall attractiveness of a location which is important. Future national innovation policy will have to increase this attractiveness, not only by encouraging individual breakthroughs, but also by supporting innovative networks, while at the same time optimising a number of other locational factors in order to facilitate leading edge markets. Policies are beginning to react to these challenges, however, there is still a considerable need for orientation in this area.

Qualitative factors and dynamic upstream- and downstream-interactions are increasingly driving R&D location decisions. Thus the motives and aims underlying the internationalisation of R&D and innovation do not relate primarily to exploiting the cost advantages of globally distributed R&D units, but emphasise more the value-added effects of transnational learning processes along the whole value-added chain (research, development, production, integration into supply chains and logistic networks, marketing/sales and services relationships). The motives for establishing R&D units abroad are very much driven by learning from technological excellence *and* lead markets *and* dynamic interactions between R&D, marketing and advanced manufacturing. The attractiveness of a national innovation system will be more and more determined by 'dynamic efficiency', the ability to support learning processes in complex system innovations, and the interaction of specific institutions (firms, R&D institutes, universities, policy administration).

R&D-intensive companies are undertaking far-reaching transformations of their R&D activities. For many of these companies, the process of internationalisation in research, product development and innovation has been accompanied by an increasingly selective focus on a very few R&D locations and the concentration of innovation activities at so-called first-class centres. A parallel trend to establish a single centre of competence per product group or technology field is also taking place. The dynamics of change depend upon an enterprise's global technology strategy, on the one hand, and upon the size and resource base of the enterprise's home country, on the other hand. As a consequence, strengthening the public research system is a necessary, but not a sufficient, condition.

Due to the different facets which must be considered if the whole complex of globalisation of R&D and technology markets should be illuminated, the conference was structured around three topics:

- Globalisation and changes in industrial relations
- International R&D strategies and centres of competence: the firm perspective
- New tasks of national innovation policies.

However, in order to meet the different needs of the readers better, we rearranged the various contributions in the following three parts:

(1) Introduction and Policy Perspective

(2) Globalisation of R&D and Technology Markets - Trends, Issues and Policy Implications

(3) The Firm's Perspective.

Following this introduction, Jürgen Rüttgers, the Federal Minister of Education, Science, Research and Technology, outlined in his opening speech five issues concerning innovation policy which have to deal with internationalisation and globalisation. Summarised, these five issues are: 1. Increasing the visibility of regional competence, 2. Merging research and application, 3. Supporting and facilitating business start-ups, 4. Opening up the German education and innovation system, and 5. Building international bridges.

Further implications for national policies (research and technology policy as well as other policy fields like social policy) resulting from globalisation generally, and from technological globalisation particularly, are the central focus of the second contribution of *Jürgen Rüttgers*, the "Petersberg Theses". Finally, he concluded: "We must thus seek to deal energetically with the challenges of globalisation. The basic ideal of solidarity-based society security must be preserved, but the system must be oriented to the future. The Federal Government is taking this approach. The aim is to preserve the social market economy - in the interest of globalisation."

At the beginning of the second part "Globalisation of R&D and Technology Markets - Trends, Issues and Policy Implications", the currently popular catchwords "globalisation" or more in keeping with the subject of the conference "technological globalisation" or "techno-globalism" must be more exactly defined, in order to proceed to a precise description of the types and extent of globalisation of R&D and technology markets. This is the starting point of the paper "Globalisation of R&D and Technology Markets - Trends, Motives, Consequences" by *Andre Jungmittag, Frieder Meyer-Krahmer* and *Guido Reger* from the Fraunhofer Institute for Systems and Innovation Research (ISI). Based on a differentiation between three types of technological globalisation, they present their chronological developments, extents and dynamics. Subsequently, the internationalisation patterns of industrial R&D in the three key technologies pharmaceuticals, semiconductor technology and tele-communications technology are described. In particular, the fact that the criteria of choice of location for R&D depend not only on supply factors such as a well-developed research infrastructure, but also that demand factors are increasingly gaining in significance in company decision-making, is focused on. Then motives and consequences of globalisation of R&D and the foreseeable development trends are examined. Also, the role of "lead markets" is presented. To conclude, the most significant consequences and starting points for a national education, research and technology policy are dealt with.

Ulrich Hiemenz, Olivier Bouin, David O'Connor and *Dominique van der Mens-brugghe* from the OECD Development Centre present a summary of the OECD study "The World in 2020: Towards a New Global Age", which was published in November 1997. Considering OECD members as well as non-members, this study explores the implications of continued globalisation for the world economy over the next 25 years in a quantitative way. Key questions were: What are the economic benefits to be produced by globalisation? Is it worth the effort? Is continued globalisation feasible and sustainable or are there going to be bottlenecks in food or energy supply? What about social cohesion in the face of accelerated structural adjustments? And, will poor countries be marginalised by international economic integration? Model-based results are provided for a high growth scenario assuming a full liberation of trade and capital flows by 2020, and a low growth scenario assuming "business as usual". A high growth scenario would result in substantially improved living standards across all regions and some convergence of non-members' per capita incomes towards OECD levels. Conversely, the world would experience much less income convergence under a low growth scenario. The study concludes that in the event that the world economy were to realise high growth over the next 25 years, an agriculture or energy problem would not be likely, but there could well be an environment problem, and widening wage inequalities in OECD labour markets could remain a challenge.

Sylvia Ostry from the Centre of International Studies, University of Toronto, describes "Globalisation Implications for Industrial Relations". She highlights that the most fundamental implication of deepening integration is the pressure for convergence of domestic policies or, indeed, of domestic systems which comprise the full array of institutional structures. Among the most significant institutional structures in domestic economies and societies are those governing industrial relations, which include not only labour-management or collective bargaining regimes, but also government policies that directly affect labour markets. The impact of globalisation on industrial relations is highly significant for a number of reasons, including firm competitiveness, which is increasingly dependent on innovation, domestic growth rates and hence job creation, and, of course, virtually all political and social dimensions of democratic countries. Thus, her paper reviews the extent and nature of convergence in industrial relation systems within the OECD during the 1980s and 1990s. It also attempts to deal with the question of the costs and benefits of divergence (working under different rules in a global economy). In doing so, the paper focuses especially on transatlantic differences in industrial relations regimes, highlighting the unique nature of the US model. Finally, the paper also briefly examines the implications of information and communication technology on labour market outcomes and on innovation systems.

Luc Soete from MERIT in his contribution "European Integration as an Answer to Technoglobalism" starts from the conclusion that national research and technology policies have to play an important role in strengthening the efficiency of R&D investment, developing the competitiveness of industry in the face of global competition and in improving the quality of life. These roles in what is generally referred to as science and technology policy are recognised and emphasised by practically every EU member country, the current US administration, Japan and most other OECD and non-OECD countries. However, Europe has had an "experimental" advantage compared to the other Triad countries in developing supra-national science and technology policies. The question addressed in his paper is whether Europe, North America or Japan will be able to learn from this experience and use it to develop effective international supra-national policies and programmes to improve the global productivity of their research and technology system, or whether the European experience has actually developed in what could be called a "first mover" disadvantage, and is in need itself of being redesigned and reoriented.

Moving from the general to the particular, the contribution of *Jürgen Mittelstraß* from the University of Constance concentrates on three points: 1. In a global economy not only economic structures change, but social structures as well. Institutional isolation dissolves. This development also affects education and research. 2. Research up to now has mainly been defined by the distinction between pure basic research, application-oriented basic research and product-oriented research. This distinction no longer holds water, i. e., all research forms have a dynamic relationship to one another. They form a research triangle. 3. When economy becomes global and research moves in a research triangle without any restrictions, (academic) education has to move out of its disciplinary boundaries. The future of research and learning, i. e. the future of (academic) education is problem-driven transdisciplinarity. He also provides all three points with brief elucidations.

The third part, "International R&D Strategies and Centres of Competence: The Firm Perspective" started with the contribution "Improving Local Conditions for Innovation - The Scandinavian Perspective" of *Yrjö Neuvo* from Nokia Mobile Phone. In his introduction, he emphasises that innovations are needed at all levels of the organisation and in all kinds of businesses. Innovations should be found, in addition to products, in production, management practices, in quality issues, or in other words, in all corporate processes. Innovation consists of four basic elements:

knowledge, hard work, excitement and enthusiasm, and knowing the goals. These elements need continuous nourishment for innovations to flourish. Therefore, he concluded, rapid interaction, readiness for change and an ability to put ideas into practice are signs of a well-functioning, innovative environment. Combining these with multinational activities creates a challenge to any company. The question of how to support best skills in every individual in order for them to be part of effective teams, which again should work effectively with other teams and even across cultural borders, is something where better solutions should always be thought of. So, advantages sought with e. g. multisite R&D may include better access to the sources of newest information from international research as well as better understanding of customer needs. This also allows easier access to qualified personnel in larger quantities. The challenges are in the creation of functional networks, division of tasks, and in preventing the undesirable side effects like the "not invented here" syndrome to appear.

On the same topic, the geographical point of view is changed in the next contribution "Developments in the Global Management of R&D - The Japanese Perspective" of *Hans-Georg Junginger* from Sony Europe GmbH. He points out that Sony's policy has always been to produce where the market is. With regard to R&D, Sony recognises that they must reflect the strengths and competencies of the host region or country. Sony's R&D presence in Europe reflects this as well. It is part of a global network of R&D operations which also supports product development in other zones of the world. The primary advantage of local R&D for Sony in Europe is that they can concentrate on developing and achieving products and technologies of European origin and ideally suited for the local market. Sony is playing a full part alongside European based electronics companies, for example, Sony has joined BMBF- and EU-funded programmes. Thus, Sony in Europe considers itself a European company as a member of the global family of Sony companies.

The contribution "Improving Local Conditions for Innovation in the Pharmaceutical Industry" of *Norbert Riedel* from Hoechst Marion Roussel Inc. switches the point of view from a geographical to a sectoral perspective. He emphasises that there are a lot of positive signals in Germany: mobilisation of venture capital, regional approaches, increasing investment in biotechnology, and the formation of alliances between US-based or international pharmaceutical companies and German biotechnology companies.

In his contribution "Policies to Strengthen Innovative Start-Ups and SMEs inGlobal Competition", *Falk Strascheg* from the Technologieholding VC GmbH and The European Venture Capital Association points to certain reasons for a divergence between Europe's technology position and its market position in the high-tech arena. From his point of view, the main reason for this gap is certainly a less developed entrepreneurial culture, a weaker venture capital industry and an investment banking community with little high tech experience compared to the US. However,

8

he also observed a lot of positive trends in Germany as well as in Europe. Thus, he concluded that improvements in the educational systems towards entrepreneurship, further support of the venture capital industry and a higher quality of the European investment banks should make it possible to have more innovative start-up companies as recognised global players in future.

Finally, I would like to express my thanks to the following persons who contributed to the success of the conference. Firstly, to Dr. Andre Jungmittag, who was the main organizer of the conference from the Fraunhofer Institute for Systems and Innovation Research (ISI) and Engelbert Beyer, who was the responsible co-ordinator from the Federal Ministry of Education, Science, Research and Technology (BMBF). Secondly, to the supporting staff, Therese Häse, Yvonne Raschke, Martina Schöntube from BMBF, and Traudl Geibel and Rainer Bierhals from ISI. Thirdly, to the BMBF, which financed this international conference and commissioned ISI to prepare and organise this event. Last but not least, thanks go to Asta Loidl and Chris Mahler-Johnstone for preparing the manuscript and the camera-ready copy of these proceedings.

Opening Speech

Federal Minister Dr. Jürgen Rüttgers

Ladies and Gentlemen,

The world is changing. I believe that historians of a later time will speak of this period of change at the end of the second millenium as a revolution.

Our age will be described as revolutionary, just as we call the events that started in England some 200 years ago Industrial Revolution from our current point of view.

What encourages me to make this claim?

- First, we are witnessing a revolution of political systems. The major powers are no longer confronting each other as hostile blocs but as variations of a general market-based order.

- Second, we are witnessing a revolution of communications. The network of global information and communication links has become unbelievably dense. The magic word of dematerialisation is doing the rounds. We are developing into a knowledge society.

- Third and last, we are witnessing a cultural revolution. With the triumphant progress of the Internet, English has finally established itself as world language and new lingua franca of the modern age. Television is broadcasting the same words and images to people in Berlin, Paris, London, Tokyo and New York at the same time.

No one has used these major changes as quickly and consistently as business. Globalization has many facets and, in many respects, a potential not nearly exhausted. So far it has been understood and described almost exclusively as an economic phenomenon.

At the same time, however, the players, that is to say, entrepreneurs and managers acting on the international stage, as well as scientists, politicians and many others, are sensing that this may only be the beginning of a development which will change the face of the world.

Hardly any other phenomenon is mentioned as frequently in speeches, analyses and justifications as is the word globalization. And yet we are nowhere near fully understanding or grasping this process intellectually.

This is the crucial reason for this congress.

Let me thank you all for your participation. I hope that we will be able to arrive at greater clarity in particular in the frequently still undefined links between globalization on the one hand and research and innovation on the other hand. Clarity that is needed by a modern innovation policy which is to open up to people the opportunities of globalization.

Not many people can, at this time, see the opportunities inherent in globalization. The feeling of being at the mercy of anonymous forces generated far from one's home region and one's own country makes people feel insecure. In a country of the size of Germany, this feeling may be more intense than, say, in the United States of America.

But I do contradict the claim that these fears and anxieties are irrational.

For the changes I have outlined are by no means uniform or unbroken. In this respect, too, they are similar to revolutionary processes.

- For as homogeneous as certain aspects of a global culture may appear to be, there are sufficient deviations and counter-movements. Major tensions result from the increase in significance of orthodox beliefs, in particular Islamic fundamentalism. It is not without reason that Samuel Huntington's thesis of a clash of cultures has met with such resonance.

- The opportunities inherent in international information and communications, the fascinating development of the Internet, coincide with counter-movements and threats. Cliffort Stoll, one of the pioneers of the Internet, has recently warned against the danger of dehumanisation in a wasteland of junk data. Political radicalism and various types of crime have apparently found a new home in computer networks. There are also fears of a new social division, a division resulting in people with computer skills, and people without such skills who form a new proletariat.

- And, finally, while it is correct that we have the privilege of witnessing the disappearance all over the world of the belligerent antagonism between capitalism and communism and the development of a global market economy, we cannot help noting that market economy and democracy are by no means natural siblings, as most of us probably believed before the iron curtain broke down. The most striking example confirming this is China.

Even in the free world, the market order is not as uniform as people may have felt it to be in the era of political blocs. In fact, it is possible to distinguish between three relatively distinct models.

- There is, first, the American model, with its great freedoms, its dynamic, its marked individualism, but at the same time its tensions and its lack of social balance.

- Second, there is - as I see it - the Asian model, in which reasons of state, discipline and a high collective readiness to perform have produced enormous success within a very short period of time. Particularly in the recent past, however, this model has also revealed its weaknesses.

- And finally, there is the European model. I would venture to draw attention in particular to our system of social market economy, also called Rhenish capitalism. Under this system, high scientific and economic performance is coupled with social balance and a - partially exaggerated - wish to achieve consensus.

- This system in particular is finding more and more followers among those who were against it only a few years ago, while its followers not infrequently feel that its core - the combination of the market economy's efficiency and social balance - has become outdated.

Very likely, there will be economic competition between these models within the global market economy for many years or even decades to come. It is my political goal to preserve the European model, in particular the system of the social market economy, in order to enable it to retain its benefits even against the background of a global market economy and to prove its great efficiency.

A division of society would in the long term smother creativity and productivity potentials, which will be so important in the future. Our answer to the challenges of globalization must be to enable both individuals and society to cope with change by education, training, research and technology.

I believe that against this background of confusing trends and discontinuities, it is more than clear that there is still a substantial need for investigations, studies, explanations and new strategies.

We will of course not be able to completely satisfy this demand here on the Petersberg. But I am confident that this congress will produce some essential and interesting results pertaining to research, development and innovation at a time of globalization.

The results of your work will immediately be submitted to Federal Chancellor Helmut Kohl's National Technology Council. I anticipate that we will get new ideas and impulses for the Federal Government's innovation policy as well as hints for strategy development in German industry and for the future orientation of the scientific community.

In particular, I hope that the intellectual discourse will be stimulated.

While in France, Great Britain and the United States, many scientists and intellectuals are participating in a public discourse, major sections of the intellectual elite of our country have hitherto refused to contribute to this debate.

On the one hand, this is understandable because the national challenge of German unification, which forms part of a world-wide upheaval, has been making major claims on Germany's intellectual resources.

On the other hand, the elite is often content with analysing changes and subsequently lamenting them.

What we need, however, is a debate identifying future opportunities and making change possible. This is the only way to open up new opportunities, rather than defending vested rights. It is the only way to take people in Germany along on the road to the knowledge society.

The astonishing changes that are coming about daily in this country show that that is possible. All we need to do is to look at the enormous adjustments made by people in the new German Länder.

However, in the West, too, more is happening than is perceived by the public. An impression of a general paralysis is created because elites are frequently refusing to participate in the necessary discourse instead of determining the goal and direction of change, because institutions frequently only tolerate changes and defend collective vested rights instead of developing new forms of reconciling conflicting interests.

This state of affairs, however, jeopardises the institutions which every society needs for cohesion. The old division of tasks, according to which the scientific community is in charge of developing knowledge in solitude and freedom, while industry is responsible for making profits, and policy-makers for ensuring justice and acceptance, this division of responsibilities does not fit the age of globalization.

Scientists making their contribution to creating new jobs do not violate the principle of the freedom of research.

Universities which train young people for specific occupational opportunities do not infringe upon the ethos of scientific rigour.

Entrepreneurs investing abroad at the same time strengthen Germany's economy, rather than weakening it.

Policy-makers wishing to make changes possible together with people while avoiding a division of society into winners and losers do not prevent change but preserve the basis of science and industry.

Now, before you begin presenting your papers and deliberations, I would like to make a political comment. My own approaches to internationalisation and globalization are probably known to some of you. I will therefore outline only five major lines.

I have five points to make, that is, five projects that are already under way.

1.   Increasing the Visibility of Regional Competence

Aside from traditional industrial locations, so-called competence centres are increasingly gaining significance. The pace of progress of certain categories of goods is increasingly determined by only a few competence centres in the world.

Thus, about 15 years ago, Silicon Valley reinvented the computer as it had previously invented the chip; today it has a major share in influencing the development of software and multimedia.

Firms wishing to keep up in this business are present in Silicon Valley, or at least follow closely what is being thought out there.

Competence centres are magnets of knowledge and pertinent skills; they attract talent and investment and at the same time determine the pace and direction of development.

Co-operation, the advantages of close contact and regional innovation cultures play a major role in the innovation process.

Believing that regional technology centres can be planned and designed by the Federal Government would clearly be a mistake.

However, it does make sense to provide stimuli for the self-organisation of competence networks. The Federal Government intends to do so increasingly in the future.

We have introduced the BioRegio competition as an instrument of developing the co-operative advantages of regions.

Since we have done so, biotechnology is unmistakably on an upward path in Germany.

German companies which years ago transferred their research and production facilities abroad are now returning to Germany to invest here. Researchers come back
also.

2.   Merging Research and Application

Yet another instrument introduced by the Federal Government are the so-called lead
projects.

This means that we intend to encourage broad-based networks of companies, research establishments, technology manufacturers and users.

I expect that the lead projects will trigger substantial system innovations in Germany and drastically shorten innovation cycles.

I am thinking of telematics systems in conurbations, the design of a closed-cycle
economy that is easy on the environment, applications of genetic engineering in
medicine and innovative multimedia applications.

The immediate responses and the quality of the proposals we have received show
that there is an enormous potential of innovation-related knowledge in Germany.

All that was needed for such knowledge to be turned into innovation was a stimulus
to start the process.

Lead projects seem to have exactly that effect: The lead-project initiative is changing reality through new thinking.

Partnerships are being formed between companies, research establishments, public
authorities - partnerships that would hardly have been created without lead projects.

The lead-project process, in addition, reveals that special services potentials are
inherent in technological and organisational innovations. This is a particularly important point.

Particularly in the case of services, what we need are not only new products but also
innovative forms of corporate organisation, innovative flow concepts, management
methods that are conducive to creativity, and the like.

In Germany, for a long time, we did not take the challenges posed by the service
sector seriously. The economic sector that is the most important for employment,

value added and competitiveness already today is the very sector where Germany is particularly weak.

Especially in the tertiary sector, world-wide competition today is relentless. Services must today be highly efficient, highly productive and of high quality if they are to be sold in the intensely competitive national and international markets.

Competition, performance transparency and quality criteria are still alien concepts in wide areas of the service sector. In the health system, the free professions, the scientific community, and the public sector, we sometimes have conditions which are strangely outdated.

It is impossible for real markets, with their pressure towards greater efficiency and productivity, to emerge under such conditions! It is considered perfectly normal in industry for productivity advances to be gained year by year. In medicine or science, such demands only meet with amazement and disbelief.

It is possible for a country to be highly innovative and strong on services, and even to export such services successfully - but it is not possible if services are mainly considered an employment policy element, and if low labour productivity is in fact even welcomed.

A prerequisite of a sound external and internal position in the service sector is that competition, productivity development, innovation and research also penetrate this sector.

3. Supporting and Facilitating Business Start-ups

Encouraging new businesses and supporting small and medium-sized companies is of central significance. The OECD has found that 5% of small and medium-sized companies have contributed almost 70% of the new jobs created in OECD member states.

The Federal Government has established the goal that nobody with commitment and ideas who is prepared to take a responsible entrepreneurial risk should be allowed to fail in Germany because of a lack of capital. We have come a lot closer to this goal.

The venture capital market for early-stage funding for innovative enterprises has recently developed at a tremendous pace in Germany.

Largely unnoticed by the general public, the German market has meanwhile become the No. 1 in Europe.

- In 1996 alone, we were able to facilitate the provision of some DM 300 million in equity participations via our programme on venture capital for small technology-based companies. This was a new record. Within a matter of two years, we witnessed an increase of 240%. And the trend continues to point steeply upwards.

- The number of biotechnology firms in Germany, for example, doubled between 1995 and 1996 from 75 to 150 companies. We may safely assume that this figure will again double in the current year, to a total of 300 firms.

- The provision of venture capital for biotechnology is booming. At present, DM 565 million in private venture capital are currently available for BioRegio activities alone. This is a development previously unknown on the German market for innovation capital.

- With the Third Financial Market Promotion Act, the Federal Government again substantially improved the regulatory framework for innovation funding. New stock market tiers are being generated.

We must not, however, stop at this. The number of innovative businesses is increasing, but not sufficiently so. Much has happened on the market for venture capital. We must now endeavour to introduce more potential entrepreneurs to this market.

4.   Opening up the German Education and Innovation System

The German education system, too, is confronting a historic challenge. Last week's student protests made this clear to everybody.

I sympathise with the protests of the students. But at the same time, I can understand that some people feel uncomfortable in the face of the chorus of solidarity addresses from very different camps.

These solidarity addresses obscure the fact that they stand for widely differing ideas about the university of the 21st century. For many of their authors, the demand for more government funding is the smallest common denominator.

It is correct that, in the face of the overload that has persisted for several years now and in the face of continuously increasing student numbers, government must make available more money to higher education institutions.

But it is also true that neither the Federal Government nor the Länder will be able to raise the necessary funds for problem-free government funding. - To say so openly is a command not only of honesty but a duty to the mostly well-intentioned students.

Basically, what we are experiencing now is an example of organised irresponsibility, which so often characterises our public debates: Everyone is involved, but nobody feels responsible.

Without any structural reforms, the university of the 21st century cannot be brought into being. The new Framework Act for Higher Education would pave the way for such structural reforms. But it is necessary to also use these new opportunities.

Hubert Markl, the President of the Max Planck Society, recently accused education policy-makers of regarding exclusive government responsibility, and consequently the mandatory funding of this whole system from public funds, as the most important guarantee of equal opportunities in education. Markl felt that, at best, this meant that nobody should have better opportunities than anyone else.

In an age of global competition, standards are no longer defined at the national level. Despite many university partnerships and trips to foreign countries, the higher education institutions have not taken up this challenge to a sufficient degree.

Hubert Markl went on to say that, all the while, the much-maligned so-called better off were increasingly sending their children to outrageously expensive foreign private education institutions, thus giving the lie to the government's egalitarian educational policy.

Foreign students and scientists are vitally important for any location. Opportunities for co-operation in the generation of knowledge as well as the markets of tomorrow are being opened up via today's study contacts.

Enhancing study offers for foreign students is a central answer to the challenge posed by globalization.

- For this reason we are supporting the creation of foreign language teaching events. This goal is also supported by a considerable number of German higher education institutions.
- We have launched a pilot programme on internationally-oriented study courses, which started this winter semester (97/98) with the first 13 study courses.
- In co-operation with the German Academic Exchange Service (DAAD) we have launched a marketing project for German higher education institutions in Indonesia. Prominent German participation in the largest Asian education fair in Jakarta served the same purpose.

- We have launched a partnership programme with the MIT. The programme enables students of this top-ranking American university to spend study periods and do practicals in Germany.

- With the proposed amendment to the Framework Act for Higher Education, we will pave the way for the awarding of internationally comparably degrees by German universities as well as a credit point system, which is to facilitate transfers between universities.

Also by improved information and systematic marketing, there is, in general, an excellent chance for German higher education institutions, in keeping with their scientific and technical potential, to become internationally attractive again for young people from all over the world.

5.　　　Building International Bridges

An active innovation policy in an era of globalization includes the internationalisation of domestic research. Within the framework of a reorientation of the German research system, we encourage the organisations sponsoring research and science in Germany to join us in pursuing this goal.

We actively support the globalization of German research. In turn we expect others to open their doors. People thinking of boundaries in connection with technology transfer limit the chances of their country.

- The Max Planck Society (MPG) took first steps in this direction already some time ago. An extension of the Max Planck Institute for Psycholinguistics in Nijmwegen was inaugurated only recently.

- The German Research Association (DFG) has traditionally maintained excellent international contacts. It is currently active in particular in China. In Peking a centre for joint research projects is being set up in co-operation with the Chinese National Science Foundation.

- Special emphasis is being placed on internationalisation by the Fraunhofer society (FhG).

- The FhG has set up research centres on laser technology, production engineering, materials research, medical technology and computer graphics in the United States. In February another centre for advanced software development will be opened in Maryland.

- With its presence in the United States, the FhG is an important partner also for German enterprises wishing to do business in the US. At the same time the FhG acts as a showcase for Germany and its science base.

- The Fraunhofer Resource Center for Medical Technology in Florida now offers an additional 30 places for practical and theoretical apprenticeship training and continuing education for technical assistants. The first three of a total of nine planned training courses started in June 1997.

- The yields from contracts from the United States rose to a total of DM 11 million during 1997. Altogether the commitment in the United States since 1993 has resulted in approximately DM 22 million's worth of contracts from the US.

- As a second leg of strategic relevance, the FhG is beginning to seek access to the Asian region. The FhG's liaison bureaus in South-East Asia and China support the FhG's institutes already active there in their marketing and acquisition efforts, which are difficult in these markets.

What is true for the commitment of German research establishments abroad conversely applies also to foreign research establishments and innovative enterprises in Germany.

Establishments of this kind from other European countries as well as from the United States and Japan have been present in Germany for a long time.

We are glad that comparable contacts are also beginning to be established with the emerging countries of South-East Asia. South Korea, for example, has set up a research institute for environmental technology in Saarbrücken.

As an answer to the trends of globalization, the Federal Ministry of Education, Science, Research and Technology intends to adjust its funding policy rules to the new challenges.

These measures include, among other things, extensive possibilities of participation by foreign companies and scientific establishments in German collaborative research networks.

This shows that German funding rules can certainly hold their own on an international basis, in particular in comparison with the much more restrictive rules applied by the US administration.

Ladies and Gentlemen,

If we wish to give people in Germany access to the opportunities of globalization, we must be open and innovative.

Ideally, we should develop a positive attitude, of the kind expressed in the words of the Indian poet and philosopher Rabindranath Tagore.

According to Tagore, "we have come into this world, not only to know it, but to affirm it."

With these thoughts, I would like to conclude and to wish you a positive and successful congress.

Thank you.

Globalisation of R&D and Technology Markets - Consequences for National Innovation Policy "Petersberg Theses"

Jürgen Rüttgers

---

1. Globalisation refers to the current economic integration that is shaping a global single market with globally operating companies; this liberalisation of the global economy is fostering greater competitive intensity and an ever-finer network of interconnections. This network is producing new growth opportunities as well as structural upheavals.

In the last decade of the 20th century, the global economy is undergoing profound changes, but it is also clearly on a growth path, caused primarily by a new stage of global economic integration: globalisation. Globalisation means more than extensive elimination of trade barriers following the Uruguay Round. The reactions of companies are of central importance in this context - especially the following:

- the direct, world-wide market presence of globally operating companies, as a result of the availability of the appropriate logistics, transport and communications technologies;

- the world-wide "expansion" of companies through rapidly increasing direct investments in foreign countries; more and more, such investments involve relocation and distribution of companies' core competencies.

Those who see globalisation as a globally exaggerated, fashionable term of the 1990s are suffering from a dangerous illusion. Every third employee of German companies in technology sectors now works in a foreign country. In many high-tech companies - such as pharmaceuticals companies - at least every second employee now works abroad. According to the United Nations' most recent calculations, the actual value of investments made by TNC's abroad is in the neighbourhood of US\$ 1.4 trillion in 1996. Estimates predict that the volume of direct investments will double by the year 2001. More and more of the global economy's value-addition process is taking place within networks of globally operating companies.

In the past decade, world trade has grown twice as fast as the gross global product; international direct investments have grown three times as fast, and international trade in stocks has grown ten times as fast. Already, production of goods and services abroad has become larger than exports, by a factor of 1.2 to 1.3.

Global economic interconnections have been facilitated by the development of information and communications technology. This is reflected in the growth of telecommunications. International telecommunications traffic - in public networks alone - more than doubled between 1990 and 1997, and it is continuing to grow exponentially. On the average, information technology costs have dropped by 30 % per year in the past decades. The price of a three-minute phone call between London and New York has dropped from $300 in 1930 to $1 today.

Companies are optimising their activities - including research, production, planning and sales - within an international perspective. In the growth sectors of the economy - where investments decide growth and jobs - decisions regarding new facilities and innovative core business are based on international comparisons of locations. Competition for investments and innovative operations is intensifying, and has become a new factor in national policy.

On the other hand, economic policy mistakes of individual countries have an immediate world-wide effect, as a result of these interconnections. This is illustrated by the impacts of the current currency and banking crisis in Asia.

---

2. The movement toward the knowledge society is significantly affecting globalization. It makes innovation a criterion for the success of companies and economies, since highly developed industrialised countries' best opportunities for competition and growth are found in new products and services that require a high degree of technological competence. Only with this emphasis will it be possible to integrate new countries entering the world market - countries from eastern Europe, Asia and Latin America.

The development toward a world-wide knowledge society is giving globalisation its real drive:

- The percentage of technologically sophisticated products in world trade is increasing rapidly. Larger and larger amounts of all goods and services involve new products and newly developed know-how.

- Knowledge is becoming a decisive factor in the economic process. Invested know-how - rather than raw materials or manufacturing costs - now accounts for most of the value of many products.

- Economic success, i.e. high incomes - is being achieved primarily with research-intensive products that succeed in the world-wide innovation competition. A recent New York Times editorial explained the U.S.' current economic success as follows: "We are the leaders in everything that is light: microelectronics, software, pharmaceuticals, computers."

The broad movement toward market economies in eastern Europe, Asia and Latin America, areas which contain over 2 billion people, requires leading industrialised countries to focus systematically on new research-intensive products and services. Economic and political transformation in central and eastern Europe; economic opening-up and liberalisation in China; India's movement toward a market economy; and extensive democratisation and market-economic orientation in Latin America have all spurred industrial growth. In most of these countries, this growth is greater than 5 % annually.

To a large degree, these countries are entering the world market on the strength of lower wage costs, which attract wage-intensive manufacturing and services. This has resulted in a loss of jobs in affected industrial sectors in Germany and western Europe.

This shift has had negative consequences for labour markets in Germany and western Europe - for example, because eastern Europe has competed successfully and palpably for jobs, but has not (not yet) become an equally significant market for German and western European products. But now demand in eastern Europe and Asia is growing rapidly and is beginning to compensate for this imbalance.

The established industrialised countries could have tried to slow the entry of these countries into the world market, by means of even greater cost reductions - but this would have been a wrong strategy. If their aim is to achieve full employment again, with high incomes, they must focus their efforts on sophisticated products and services that will be in demand in the new growth regions of the rapidly growing world market.

> **3. The new and much-discussed strategies of internationally operating companies are a logical response to the changed competitive situation as a result of globalisation; world-wide competition among locations, for investments, is becoming a significant second tier of competition; the degree to which German companies are internationalising is in line with the world-wide trend.**

The following features of global corporations' strategies are particularly significant in light of globalisation:

- World-wide shifting of locations, in which individual locations are often set up as competence centres with their own R&D and market strategies.

- World-wide relocation and networking of research; companies are gravitating to areas in which new product families and markets are developed, in order to sharpen their technology and market competence.

- In general, a sharp "functionalisation" of R&D: i.e. downsizing of central research departments; greater assignment to segments; internal use of the customer-contractor principle; where possible, outsourcing of research, networking and co-operation strategies to other companies and institutions.

- Concentration on the company's core competencies, the areas seen as likely to succeed in the world market - with the appropriate streamlining of the company's activities.

- Flexibilisation, through reductions of the depth of production, and increased outsourcing of supply and services.

- The last of these strategies has an impact on the supplier sector and on industrial services:

- On the one hand, it creates room for new entrepreneurial profiles - including company-oriented services.

- On the other hand, systematic outsourcing shifts research and development risks to the supplier sector. Intense competitive selection drives a process of concentration and specialisation, and encourages companies - including small and medium-sized companies - to focus heavily on innovation.

The internationalisation of German companies within this framework has developed rapidly in recent years, and it is a logical response to the challenge of globalisation: German companies' direct investments in foreign countries have reached a level of 1% of the country's gross domestic product - the same figure found in the U.S., Japan and France. The comparable figure in the UK is 2.5 %. German companies are part of the trend toward international networks. It thus certainly cannot be said that Germany is a "world champion in exporting jobs". There is cause for concern, however, in the facts that foreign companies are investing too little in new production capacities in Germany, even though companies are continuing to pump income into their already large volume of foreign investments.

Without doubt, cost plays an important role in the competition between locations. On the other hand, German companies' direct investments in the U.S. are far greater than their investments in countries with dramatic cost differentials. The central motives for these investment decisions include market strategies, the issue of competence expansion and the need to be present and obtain local advantages in new, rapidly growing markets.

> **4. National and regional educational, research and innovation systems are becoming countries' most important capital; they are also becoming strategic factors in economic policy; know-how relative to new markets is becoming a central resource, one that influences selection of company locations; the position of German companies in the innovation competition has improved.**

In innovative sectors - which have a strong influence on future jobs and prosperity - leading companies are striving to be present in areas with the best conditions for knowledge generation, innovation and growth relative to the products and technologies concerned - often, innovative core business is combined in such areas. Top locations feature a large degree of scientific and technical innovation, and attractive conditions for production and pioneer applications. For certain products, lead markets and competence centres form that exert world-wide influence. The world's high-wage locations are having increasing success with such concentration and performance leadership.

Innovation policy that seeks to respond to these requirements of global competition must have a broad perspective, must be continually adapted and must focus on at least the following four areas:

- Enhancement of "human capital", i.e. education and training;
- The research infrastructure and research system ("research landscape");
- Knowledge management, i.e. especially the establishment of connections between research and companies, and between research and practical applications;
- A vital structure of companies that are open to international competition, and in which innovation is considered a central strategy.

Modern innovation policy must continually examine these sectors of the economic "innovation system" to see whether they themselves, the framework that influences them, or their interactions can be improved.

The following insight is centrally important: overall, the interactions between science and business do not control themselves in an adequate way. The different sectors are different subsystems of the modern society, and they are subject to different control mechanisms; i.e. they do not interact "by themselves" - at least not adequately. Consequently, one of the primary tasks of innovation policy is to achieve effective interaction - that which does not occur by itself - between knowledge and application.

German research policy of recent years has thus strongly emphasised connection and co-operation between science and business/industry. Important concepts in this connection include:

- More competition, in research establishments, for projects;
- Putting strategic lead projects to tender;
- Research co-operation among small and medium-sized companies;
- Creating connections between research establishments, industry and administrations on the regional level - as has been done successfully in the BioRegio competition.

Very significantly, awareness of the importance of innovation has been reawakened in German industry, especially among small and medium-sized companies. German research institutions' income from industry is growing rapidly. The industrial income of the Fraunhofer Society, which maintains working contacts with some 2,500 companies, has been growing especially rapidly. Since 1995, the BMBF's research co-operation programme has awarded contracts for over 3,000 projects to small and medium-sized companies. On the whole, the conditions for innovation competition in Germany have been considerably improved in recent years, in both research and technology.

Recent figures indicate that investments in innovation have been growing considerably in Germany. Growth and export successes have been driven by research-intensive industries. New areas of competence have been established in laser technology and robotics. German companies are at the forefront in production and motor-vehicle technology. There is optimism in the biotechnology sector. The emphases in microelectronics have developed in areas where close ties can be established to German industry's traditional strengths - such as telecommunications, automobile electronics and production technology. Recent studies have shown that structural changes, at least in industry-oriented services, are taking place very rapidly when compared internationally, contradicting the oft-expressed pessimism in this area. A new trend toward more innovation at home locations has been able to take off.

---

5. **The keys to innovation policy include: establishment of competence centres, knowledge management in networks and more competition in research and technology-oriented start-up companies; these areas must be strengthened.**

Major areas of scientific-technical development are focused in regional competence centres. Some 15 years ago, for example, Silicon Valley redefined the computer; previously, it invented the microchip. Today, much software and multimedia development begins there. Companies that wish to compete in these businesses are either present there or closely follow that area's intellectual trends. Competence centres are magnets for knowledge; they attract talent and investments and they determine the pace and direction of development. The development of Silicon Valley has of course been unique. And yet concentrations of high-technologies are seen in many other areas - and also in Germany. Germany's federal system, and its historical development, have given Germany a polycentric system with large regional clusters of competence. The development of these clusters shows that co-operation, management advantages and regional innovation cultures play a considerable role in innovation dynamics. The growth of high-tech centres promotes employment and growth in all parts of our country.

Competence centres cannot be planned on the drawing board. They arise wherever scientific excellence and practical vision meet. The Federal Government can provide incentives for self-organisation, and has done so successfully with the "BioRegio" competition. Biotechnology is now on the upswing in Germany. The competition has shown that regional innovation strategies that bring together competence from the areas of science, industry and administration have excellent chances to succeed. One of innovation policy's increasingly important responses to globalisation will be support for regional alliances.

Throughout the world, growth centres feature great entrepreneurial dynamism. One of the priorities of policies of recent years has thus been to improve the framework for venture capital. On the strength of new forms of venture capital financing, the number of new companies in the BMBF's programme "Direct-investment capital for small technology-based companies" (Beteiligungskapital für kleine Technologieunternehmen) is growing exponentially. In 1996 alone, this programme provided some DM 300 million worth of venture capital. An increase of 240 percent was achieved within two years. And the trend points steeply upward. Provision of venture capital for biotechnology is booming. The "BioRegio" programme activities are currently providing DM 565 million of private venture capital.

The German venture capital market for early-phase financing has become number 1 in Europe. In a few years, this market, adjusted for size, could equal that of the U.S. New stock market segments are being established. The number of innovative start-up companies is growing. In 1996, the old Länder registered some 15 % more new high-technology companies, some 40 % more new advanced-technology companies and some 55 % more new innovative services companies than were founded in 1992.
The Third Financial Market Promotion Act (Finanzmarktförderungsetz) has considerably improved the tax framework for financing of innovation. The Federal Government's tax-reform bill provides for continued support for young technology companies. A great deal of potential for growth in the area of new companies remains. More potential company founders must be encouraged, before they turn to

the capital market, and more need to be given the necessary entrepreneurial training. A decisive role is played in this connection by initiatives for more training in the area of self-employment - including training at higher education centres for company founders (Gründerhochschulen). Opportunities are also provided by special informal markets, similar to those in the U.S. (Business Angels): in such markets, private persons provide capital, advising and management capacities for high-growth companies. The positive statistics on start-up companies, and the numerous private support arrangements, show that a new culture of start-up companies is possible in Germany.

---

6. **Research and development must be capable and usefully organised; in addition, the economy requires the proper legal framework for pioneering applications and start-up companies; this is another area in which Germany's position has been considerably improved.**

Trends in companies' R&D investments are strongly affected by market forces and market dynamics. Lead-market functions occur wherever systematic, concentrated innovation takes place and matures in close contact with demanding, leading, and innovative customers. Around the world, ways are being discussed to improve the framework for such dynamic processes.

The solution includes lower taxes and fees, greater tax equitability and a "lean state", greater flexibility in the labour market and enhanced entrepreneurial independence. Policies for improvement of such so-called "hard" location factors will remain important in the coming years.

In recent years, considerable success has been achieved in Germany through deregulation of administrative procedures, of labour laws and of business hours; as well as through privatisation of the postal, railway and telecommunications sectors. This progress must be continued.

Internationally exemplary regulation and certification standards can have a decisive impact on lead markets. On the other hand, inappropriate national regulations will be irrelevant for developments being pursued globally - they harm only the region to which they apply. For example, amendment of the Genetic Engineering Act (Gentechnikgesetz), along with subsequent revision of pertinent ordinances, has significant improved Germany's legal framework in the area of genetic engineering.

On 1 August 1997, the Information and Communication Services Act (IuKDG) created a clear, reliable framework, with a neutral effect on competition, for the important, growing information and communications technology industries. This has enabled Germany to be a pacesetter that is helping to drive development of "electronic marketplaces". U.S. companies' investments in Germany's mobile communications sector prove that a lead-market function has already been achieved in this sector.

The legal framework for operation authorisations has been considerably simplified and now includes important exceptions for innovation and pilot production operations.

In the next few years, the BMBF plans to carry out benchmarking of the legal frameworks for all innovative sectors, using international comparisons. There is considerable need for action, for example, in the pharmaceutical sector, where clinical research suffers from inadequate co-operation between companies and universities/clinics and is hampered by licensing procedures.

Pioneering applications are created in networks of technology producers and users. The concept behind the BMBF's lead projects is in keeping with this insight. The lead projects combine sophisticated objectives with concrete perspectives for application. They also seek to promote interaction between a very broad range of different disciplines and applications.

The "lead projects competition" initiative invites research alliances to submit project proposals in defined research areas. The competition fosters partnerships - between companies, research institutions and authorities - that would hardly come about without lead projects. As the examples of new mobility concepts show: the current focus is no longer on categories of individual modes of transportation or infrastructures, but on intelligent, integrated mobility services. Lead projects consist of practical projects and pilot tests that support the search for new, socially acceptable, technical and economic markets. They can give impetus, within international competition, to new lead markets in Germany and Europe.

> 7. **Competitiveness, in the framework of globalisation, means openness and international co-operation - also in research - and not isolation or "techno-mercantilism"; an efficient Fifth RTD Framework Programme will be an important step forward for Europe.**

Science has always been international - and today, there can be no research excellence without global co-operation. Dynamic scientific, cultural and economic developments arise in centres of communication. Personal contacts remain the key to such communication - even in an era of the Internet and videoconferences. At the same time, it is becoming more and more important to be able to gather, assess and apply (if necessary) any globally generated knowledge. Important science and technology projects are hardly possible without co-operation. Internationalisation of training, openness of locations and international co-operation are significant responses to the global knowledge society.

It is important for research institutions to have broad international connections and involvement. The Fraunhofer Society is one of the main features on the German research landscape. In a unique way, it builds bridges between the worlds of science and industry. The Fraunhofer Society has always earned parts of its income through projects for foreign customers, and it has enhanced its international orientation through service and contact offices. Increasingly, it is establishing branch labs of its institutes in foreign countries, using basic government funding. Such moves have been carefully considered. An institution that wishes to sell top-quality international research must be present in the world's leading research centres and must meet the demands of sophisticated contracting entities in those areas. The home and host locations alike profit from such international involvement by a powerful research institution.

R&D capacities of foreign companies in Germany are an important element of the German innovation system. Through their intensive R&D and innovation, through their exchanges of state-of-the-art technologies throughout their world-wide corporate groups, and through their co-operation with universities and research institutions in foreign countries, such companies contribute significantly to generation of knowledge and transfer of knowledge to Germany. Companies that produce in Germany have higher-than-average investments in R&D. They tend to be active in areas in which Germany, when compared internationally, has not previously enjoyed specialisation advantages: for example, in microelectronics and telecommunications. The R&D capacities of foreign companies in Germany also balance with R&D capacities of German companies in foreign countries. According to the most recent figures, both contribute about 16 % of total R&D expenditures in Germany. German companies will further internationalise their R&D in the coming years. National innovation policy also seeks to attract large volumes of foreign R&D investments. It goes without saying that investors enjoy full access to a potent R&D infrastructure and to public R&D programmes. On the other hand, marketing of Germany as a research location must be improved over past efforts.

In Europe, we are now intensively discussing the orientation and structure of the 5th framework programme. We consider it necessary for support to be concentrated on innovative research areas that are of decisive importance for Europe. We cannot provide support for all aspects of conventional areas. We are considering a programme that would concentrate on the following five programme lines:

- Life sciences;
- Environment;
- Energy;
- Information society;
- Production and transport technology.

We consider it important to try to concentrate competence - throughout Europe, wherever a useful basis is available - and to seek to link such "crystallisation points" to form European competence networks. European support, in conjunction with national investments, must foster formation of regional focuses - so that European research does not remain at a "subcritical" level, but instead helps to determine the pace of progress.

The following important point should also be emphasised: the more important that international exchanges of knowledge become with regard to innovation, the more important it also becomes to protect intellectual property rights. For this reason, we have launched a patent initiative in Germany aimed especially at helping small and medium-sized companies protect their know-how under patent law. The necessary rights cannot be in place without such codification of knowledge. Respect for such rights does not hamper international co-operation; on the contrary, it facilitates such co-operation, because it facilitates a balance of economic interests.

---

8. **Education and training provide competitive advantages and are the best insurance against unemployment. Priority must be given to modernising vocational training and reforming institutions of higher education.**

The growing global interconnections between many areas of life are presenting enormous challenges for learning. The concept of globalisation engenders fears of "monocultures" and of "mixed cultures" and fears of the difficulty of individualisation in the face of global markets. In fact, protecting a culture means developing it, opening it. Countries and regions with a vital, open and strong cultural identity profit from such interaction. The basis for such interaction must be provided through education.

Innovation-driven changes in labour markets present new challenges for the educational system: in the last 30 years, globalisation and the emergence of the knowledge society have spurred employment of highly qualified people in all developed economies. In Germany, this trend toward higher qualifications and knowledge intensification has affected all areas of the economy. Development of innovative services can be expected to further accelerate this process. Since the end of the 1970s, the share of university graduates in the labour force has more than doubled; the percentage of employed people with vocational qualifications is increasing. New added value is created through knowledge. Similarly, the role of education in developing new employment areas is also growing.

In the past weeks and months, intensive debate about our educational sector has ensued. This debate is a centrally important discussion about the future. We need an educational sector that promotes achievement and helps everyone reach his or her maximum potential. Change is urgently required. The question regarding the winners and losers of globalisation is also a question about education and qualifications. In recent years, the labour-market position of unskilled workers has dramatically worsened in all industrialised countries. Between 1991 and 1995, 1 million jobs that required no formal vocational qualification were lost. The development of the knowledge society and globalisation are both spurring this trend: much faster economic growth is needed to create new employment for unskilled workers than is needed to create employment for people with higher qualifications. Young people with learning and achievement problems thus need special support and specially designed forms of training - for example, training concentrated on practical subjects.

On the whole, vocational training - through the dual system's great integrative power - has made integration of young people into the labour market more effective in Germany than in other European countries. Such training helps maintain Germany's advantages by providing highly qualified specialised personnel. Changes in technology and the world of work and the growing numbers of young people in training classes, are creating powerful pressures for reform. We have combined our responses to this need in the "vocational training reform project". This project is emphasising further reorganisation of vocational profiles, along with a redesign of the training framework, in order to make training more useful for companies and more practically oriented. The pace of innovation and of emergence of new qualification requirements will continue to accelerate. Consequently, in the modern economy, vocational training must be seen as a gateway to learning that continues throughout one's entire working life. The term "lifelong learning" is used to denote new requirements in this area. Support of young employed people who are eager to learn remains a central task in protecting our future.

---

9.   Exchanges of students and scientists are of central importance as a result of global interconnections. Around the world, markets for educational services are forming, especially for universities' services. These markets must not exclude Germany. Participation in them will not be possible without reforms, proper organisation and a sense of service, however.

Students' contacts often become lifelong contacts; often they open opportunities for co-operation and tap the markets of tomorrow. For decades, the number of foreign students in Germany has been growing only slowly, however. The number of foreign students in the U.S., especially Asian students, has skyrocketed over the past 10-15 years.

For this reason, one of the Federal Government's central aims is to enhance the attractiveness of studies in Germany. The reform of the Framework Act for Higher Education (Hochschulrahmengesetz) will improve German universities' chances in this international competition - through deregulation, differentiation of studies, better organisation and qualifications and an emphasis on performance. Universities and Fachhochschulen will be permitted to award the "bachelor's" and "master's" higher education degrees, which are recognised world-wide. New types of international programmes have been established that are oriented to the needs of foreign students.

The areas that need considerable improvement include: presentation of the German higher education sector abroad, the German higher education system's vocational orientation and the system's "customer orientation". For example, very little attention is given to contacts for creation of university networks of personal relationships - contacts that last far beyond student days. Research institutions and universities make very little use of opportunities for image advertising.

In the coming years, world-wide markets for educational services will emerge as societies become knowledge societies. Institutions of higher education hold fifth place on the list of the U.S.' most important export products in the services sector. The number of jobs directly tied to these exports is estimated as being far greater than 100,000. Germany has virtually no participation in these markets. One way of remedying this situation could be to develop comprehensive service packages for foreign students that would be privately financed but make use of public infrastructures.

The importance of foreign-language skills - and, thus, of student exchanges and school co-operation - continues to grow. New models must be considered in this area - for example, for foreign-language instruction in primary schools. Opportunities for training in foreign countries, during higher education and during vocational education, need to be improved. International study programmes also benefit German students. A wide range of instruments - provided by agencies and a wide array of programmes - is available to support such efforts. Priority must be placed on preserving and improving these instruments.

---

10. Accepting the challenge of globalisation does not mean creating divisions in society - it means making society more flexible; this challenge requires streamlining and modernisation of social security systems - but not their elimination.

The debate on globalisation's consequences for the social state features a considerable range of different assessments: one side paints fears of "globalization traps", of societies with few winners and many losers, of the "end of work" and the end of the social state. New international regulations are demanded to ease the brunt of the market forces.

The other side welcomes the "weakening of the state", or at least "the end of the social state" and its pressure on labour costs. Such views often give little consideration to the losers in such major changes. The discussion cannot afford to ignore the fact that necessary higher vocational flexibility must be backed up by properly adjusted social security, with input solidarity.

Making use of the opportunities of globalisation means combining such opportunities with internal solidarity. In light of the facts that the public sector accounts for over 50 % of GDP, and that social expenditures account for over 33 % of GDP, it would be absurd to speak of the end of the social state. On the other hand, system reform aimed at efficiency and a sense of personal responsibility is required. A willingness to recognise that globalisation will create winners and losers is also required, however. For the foreseeable future, the picture for income from capital will be brighter than for income from work. Consequently, new models of wealth-building are required. Unions and employers have a special responsibility in this area.

On the whole, only a powerful, innovative economy creates room for social services. We must thus seek to deal energetically with the challenges of globalisation. The basic ideal of solidarity-based social security must be preserved, but the system must be oriented to the future. The Federal Government is taking this approach. The aim is to preserve the social market economy - in the interest of globalisation.

2. Globalisation of R&D and Technology Markets - Trends, Issues and Policy Implications

Globalisation of R&D and Technology Markets - Trends, Motives, Consequences

Andre Jungmittag, Frieder Meyer-Krahmer, Guido Reger

1. Introduction

The growing integration of many countries in a world-wide division of labour, the decline of trade barriers and national regulations, as well as an increasing mobility of production factors are described today in public and scientific discussions as „globalisation", in order to pinpoint a new quality in the internationalisation of the economy. On this general level the globalisation of industry encompasses the cross-border operations of enterprises in organising their research and development, production, in the procurement of materials and services, and their marketing and corporate finance activities.[1] The distribution of various corporate activities among different countries is regarded as a crucial characteristic - or a new quality. This particular circumstance distinguishes globalisation from the previous form of internationalisation, in which a company was active in different countries, but could still be unmistakably assigned to one state of origin.[2] Today, it is increasingly difficult to pin a national label on internationally (globally) active companies. Traditionally, the international expansion of enterprises took place in the field of foreign trade, followed under certain conditions by direct investments abroad.[3] This internationalisation of production through direct investments has greatly increased since the 80s. Parallel to this, the traditional modi operandi adopted to carry out research and development (R&D) have changed and the generation of technological innovations is being increasingly influenced by the generally acknowledged trend towards globalisation. Multinational enterprises are now, after sales and production, also

---

[1] Cf. OECD (1996), p. 9.

[2] The previous forms of internationalisation are reflected also in the classical definition of the term multinational enterprise, which was probably introduced by Lilienthal (1960), (according to Aharoni, 1971, p. 27). Here it is proposed to define multinational enterprises as „corporations which have their home in one country but operate and live under the laws and customs of other countries as well" (Lilienthal, 1960, p. 119).

[3] The firm-specific advantages and advantages due to internalisation of the company directly investing counted initially as the most important pre-conditions (cf. Hymer, 1960 and Rugman, 1980). In Dunning (1977, 1979, 1981) these pre-conditions were taken up, complemented by the advantages of location and condensed into an „eclectic" approach. Not only the approaches which are based on firm-specific advantages and advantages due to internalisation, but also the „eclectic" approach can explain on the whole only substitutional relationships between exports and direct investments. Modifications to these theoretical approaches however suceed in creating complementary relations between these alternative possibilities of supplying foreign markets. Jungmittag (1996), pp. 44 - 133 provides an overview of the various theoretical approaches to explain the relationship between foreign direct investments and exports.

internationalising their R&D. What significance and what consequences this development will have, however, is and remains controversial.

In the media the increasingly international generation, transfer and diffusion of technologies is described by the catchword „technological globalisation" or „techno-globalism", which has also been adopted in the scientific field. In order to invest this somewhat ambiguous term with real meaning beyond a general catchword and to contribute to the description of the globalisation of R&D and technology markets, it must be more exactly defined. A definition of this kind (or taxonomy) should differentiate between at least three processes:[4]

- The *international (global) exploitation* of technologies developed on a national level: the companies try to exploit their technologies internationally, by means of exports, production abroad or licensing agreements. No new developments are concerned here, but their importance still increases.

- The *international (global) technological collaboration* of partners in more than one country for the development of know-how and innovations, whereby each partner retains his own institutional identity and ownership relationships remain unaltered: these co-operations can take place not only between companies (e.g. via joint R&D projects, the exchange of technical information, joint ventures or strategic alliances), but also by means of joint scientific projects and the exchange of scientists or students. Typical actors here are national and multinational enterprises and universities and public R&D institutions. Forms of international technological collaboration are gaining in significance and are also supported in the political field by corresponding programmes.

- The *international (global) generation of technologies* is carried out by multinational enterprises, which develop R&D strategies to create innovations across borders by building up research networks. R&D and innovation activities which are carried out simultaneously in the home and host country, the purchase of foreign R&D facilities and the establishment of new R&D institutions in the host countries are all means to this end. There are a number of empirical proofs that these activities are gaining in significance in a number of industrial sectors, at least for large enterprises. The picture however is not unambiguous.

A further possibility is the „global sourcing" of technologies through foreign trade (import of high-tech and leading-edge technology goods). This is certainly an expression of the internationalisation of technology markets, but is not connected to the internationalisation of R&D.

The three different processes mentioned above contribute also to analytical clarity, because the extent of their effects can be described by various indicators. The eco-

---

4 Cf. Archibugi/Michie (1995).

nomic equivalent of the global exploitation of technologies developed on a national level are first of all the foreign trade flows. They again are connected to patent applications on the foreign markets.[5] In addition to these are direct investments to establish branches abroad, which exclusively serve the value-added chain following on from R&D. The extent of global technological co-operation is reflected in international joint ventures in the corporate field, which once again can be shown in the number of corresponding co-operation agreements. In academic and public research institutions the international scientific exchange can be measured by the number of transnational co-authorships. The approximate measurement of the global generation of technologies is somewhat more difficult, due to the data situation. Direct investments in R&D facilities are first of all necessary, either in the form of buying up already existing R&D facilities or the setting-up of new ones in the host countries. The R&D output can then be approximately ascertained through the patent applications of enterprises under foreign control.[6]

The global trends observed using these indicators will be outlined in the first part of the next chapter of this paper. The findings gained on the macroeconomic level will be supplemented in the second part by detailed results from a new study „Internationalisation of Industrial R&D in Selected Technology Fields", which was carried out by ISI, DIW and ZEW (1997) for the BMBF. The third part is devoted to the motives for, and consequences of, the globalisation of R&D. Finally, the consequences for national innovation policy will be discussed. The significance of „lead markets" will play a central role therein. Necessary and possible measures for innovation policy will be pointed out.

2. Internationalisation of R&D and Innovations - An Empirical Assessment of the Current Situation

2.1 Global Trends

In accordance with the proposed division into three parts, first of all the international exploitation of technologies or innovations developed on a national level should be regarded. It is obvious that enterprises with a strong export orientation also try to exploit their technological advantages (as a reflection of firm-specific advantages) which express themselves in their innovations internationally. Also, technology-intensive goods are particularly tradable/marketable inter-nationally. Empirical studies for instance show that often close relationships exist between the

---

5 Cf. Archibugi/Michie (1995), p. 125 and p. 137.

6 The indicators proposed here are based on Archibugi/Michie (1995), who have collated evidence from numerous other studies.

development of technical and foreign trade specialisation patterns.7 These relationships can be proved in time series and also cross-section analyses. Besides, new
technologies can also be profitably exploited in foreign markets through licence
agreements, independent of goods exports. Thus domestic technological capabilities
(advantages) lead to improved export possibilities, which again give rise to intensified attempts to exploit technological innovations, embodied in goods or not, inter-
nationally.8 A further step is the setting-up or purchase of production plants abroad.

It is, however, always obligatory that a company covers the international exploitation of innovations developed on a national level by applying for patents abroad.
Patents are closely bound up with the pursuit of commercial strategies as a matter of
principle. Just because patents must be utilisable for industrial applications and because patent protection is relatively expensive, patent registrations abroad are an
appropriate indicator for strategies which are aimed at foreign markets.9

Two perspectives can be chosen. On the one hand the patent applications of the
companies in the relevant foreign markets can be observed (see *Table 1*). It can be
seen here that in all the OECD countries surveyed, the patent applications of foreign
companies increased between 1989 and 1994, in part massively.10 The domestic
patent applications on the other hand increased more moderately in almost all cases,
or even decreased. Exceptions here are the USA, where the domestic patent applications between 1989 and 1994 increased more than the foreign ones, and Germany
and Portugal, where the growth rates between 1989 and 1994 for both groups remained nearly constant. This development is reflected also in the changes in the
proportion of patent applications which fall to domestic and foreign applicants.

---

7   Cf. Münt (1996).

8   Cf. Archibugi/Michie (1995), p. 125.

9   Cf. Schmoch (1996), p. 225.

10 The term „foreign" is used here in the sense of the usual definition of the balance of payments
    statistics, where „foreign" is taken to mean all natural and legal entities with place of residence
    or usual place of residence or domicile abroad. By analogy, „domestic" refers to all natural and
    legal entities with place of residence or usual place of residence or domicile in the home country.

Table 1: Patent Applications of Domestic and Foreign Companies in OECD Countries

	Share of domestic patent applications in %		Absolute change of domestic patent applications in %	Share of patent applications by foreigners in %		Absolute change of patent applications by foreigners in %
	1989	1994	1989-1994	1989	1994	1989-1994
Australia	26,80	24,58	31,87	73,20	75,42	48,16
Belgium	2,35	1,73	-3,91	97,65	98,27	31,56
Denmark	10,49	2,91	15,54	89,51	97,09	351,12
Germany	35,57	35,55	16,66	64,43	64,45	16,72
Finland	18,09	12,09	18,94	81,91	87,91	91,08
France	17,70	15,32	-0,98	82,30	84,68	17,76
Greece	2,60	0,02	-97,77	97,40	99,98	169,87
Great Britain	23,51	19,88	-7,36	76,49	80,12	14,73
Ireland	17,40	1,98	11,41	82,60	98,02	1064,28
Italy	n.a.	12,00	n.a.	n.a.	88,00	n.a.
Japan	88,84	86,38	0,63	11,16	13,62	26,37
Canada	8,64	6,19	-16,63	91,36	93,81	19,52
New Zealand	18,07	7,38	56,26	81,93	92,62	332,19
Netherlands	6,18	3,42	-33,86	93,82	96,58	23,15
Norway	10,15	5,71	3,05	89,85	94,29	92,22
Austria	6,11	4,16	-7,29	93,89	95,84	38,97
Portugal	2,53	0,25	22,09	97,47	99,75	1155,24
Sweden	7,73	7,72	25,04	92,27	92,28	25,19
Switzerland	9,28	6,41	-12,92	90,72	93,59	30,04
Spain	6,92	4,02	2,50	93,08	95,98	82,06
USA	51,14	51,94	30,56	48,86	48,06	26,45

Source: OECD (1997), Main Science and Technology Indicators / own calculations

42

A large, almost constantly growing proportion of the patent applications in the OECD countries surveyed was filed by foreigners. Even in the USA, which has great technological potential, this share amounted to 48.86% in 1989 and 48.06% in 1994. Then Germany follows in second place, with a share of 64.45% and Australia in third, with a share of 75.42% of patents applied for by foreigners in 1994. In all other OECD countries surveyed the share of patents applied for by foreigners was over 80%.[11] Japan continues to be an exception. In 1994 only 13.62% of the patent applications were made by foreigners. This is certainly caused by the institutional nature of the Japanese patent system, but at the same time shows that western companies are relatively scarcely represented in the Japanese markets.

Table 2: Relationship of Foreign Patent Applications by Domestic Companies to the Domestic Patent Applications by Domestic Companies in OECD Countries 1989 - 1994

	1989	1990	1991	1992	1993	1994
Australia	2,15	2,58	3,18	3,39	4,03	5,84
Belgium	8,28	8,89	9,43	11,64	14,99	16,23
Denmark	7,08	9,14	11,44	14,16	16,67	24,42
Germany	4,20	4,93	4,64	4,95	5,02	5,54
Finland	3,82	5,01	6,43	6,07	11,37	13,18
France	4,44	5,25	5,01	5,48	5,71	6,40
Greece	0,87	1,33	1,79	--	2,43	3,02
Great Britain	3,06	4,03	4,42	5,06	6,39	8,49
Ireland	1,30	1,67	1,69	2,62	5,29	5,83
Italy	--	--	--	--	4,47	5,30
Japan	0,37	0,41	0,40	0,39	0,38	0,42
Canada	3,43	6,23	7,97	11,47	10,54	11,69
New Zealand	0,94	0,92	0,50	0,69	7,23	7,99
Netherlands	8,50	9,49	11,12	19,90	20,91	28,29
Norway	4,02	5,17	6,58	8,85	9,28	11,68
Austria	3,65	4,18	4,48	5,51	5,59	6,14
Portugal	0,83	1,00	0,65	0,88	5,93	5,82
Sweden	6,61	8,02	8,98	10,07	12,56	14,39
Switzerland	8,06	8,81	8,87	11,81	12,41	12,52
USA	3,19	3,58	3,59	4,69	5,39	6,28

Source: OECD (1997), Main Science and Technology Indicators

On the other hand the patents applied for by domestic companies at home and abroad can be related to each other. A quasi diffusion ratio of this type indicates how extensively patents are exploited abroad on average.

[11] The patents applied for and registered by the European Patent Office provide further evidence for the outlined development. In 1996 50.66% of the patents applied for and 52.05% of the patents registered stemmed from non-member states (principally the USA with 29.15% or 25.28% and Japan with 17.74% and 23.96% respectively. (Cf. European Patent Office, 1996).

The corresponding rates for the OECD countries surveyed are given in *Table 2*. In almost all countries a significant increase in the exploitation of patents abroad can be noted between 1989 and 1994. The exception once again is Japan, which stagnates on an apparently low level. In fact, however, this low rate is occasioned by the very high number of domestic patent applications to the Japanese patent office. Especially smaller countries with a high technological dynamism have further consolidated their already high exploitation of inventions abroad. These countries are dependent to a large extent on international markets, because of the limited capacity of their national markets. But also some peripheral countries (Greece, Ireland, Portugal and Spain) which show relatively low domestic patent activities in the domestic patent offices, have started constantly increasing their diffusion rates.

On the whole, the developments of the selected indicators determined for the first half of the 90s - with the exception of the strongly increasing diffusion rates for the periphery countries - are not new. They were already observed in the 80s.[12] All things considered, they show that the international exploitation of innovations has increased massively in the first half of the 90s.

Next to foreign trade and licensing agreements the building-up and/or take-over of production plants abroad presents a further possibility to reap international benefits from technological advantages. Activities of this kind are reflected in increases in direct investment holdings abroad, such as can be observed for industrialised countries above all in the second half of the 80s, with growth rates of 20% per annum and after a slight slump during the world-wide recession at the beginning of the 90s from 1995 - joined to a boom in company take-overs.[13] Direct investment data however do not provide information on whether really only production plants (quasi extended workbenches) were established or whether also R&D facilities were taken over during acquisitions or were set up in new foundations. A look at the sectoral distribution of the direct investment holdings would seem however to suggest that not only German companies abroad but also foreign enterprises in Germany are engaged in above-average R&D-intensive fields. *Table 3* shows the shares of German companies abroad in the total turnovers and number of employees of German companies. In a general increase in the degree of internationalisation, significant differences are apparent between the sectors. The firms in the chemical industry made 50.7% of their turnover in 1994 already in foreign branches and employed 46.6% of their personnel abroad. In the motor vehicle sector the share of turnover was 36.5% in 1994 and the share of employees abroad 37.5%. In the electrical engineering sector both shares were around 30%. A similar picture - even if not quite

---

12 Cf. Archibugi/Michie (1995), pp. 126 - 127.

13 Cf. UNCTAD (1996).

Table 3: Share of Turnover and Employees of German Enterprises Abroad in (total) German Enterprises 1980, 1990 and 1994

Branch	Turnover abroad			Employees Abroad		
	1980[1]	1990[1]	1994	1980[1]	1990[1]	1994
	Share in %			Share in %		
Manufacturing industry thereof:	16,4	21,2	25,0	17,1	21,0	23,9
Non-R&D-intensive sectors	9,2[2]	10,6	11,7	9,6[2]	12,3	14,5
R&D-intensive sectors thereof:	26,8[2]	29,2	35,1	24,1[2]	27,4	31,1
Chemical industry	38,3	48,1	50,7	42,1	45,8	46,6
Mechanical engineering	12,5	15,4	19,2	13,1	15,3	17,7
Office machinery/ADP	29,2	12,4	41,8	24,4	8,6	23,1
Road vehicle constr.	22,1	27,4	36,5	25,2	30,9	37,1
Aerospace industry	n.v.	35,9	57,9	n.v.	4,4	22,4
Electrical engineering	23,0	26,8	29,2	22,7	27,1	30,3
Precision engineering, optics, clock-making	18,9	18,7	24,4	15,7	18,1	22,2

1) Former Federal Republic of Germany (pre 1990)
2) 1982
Sources: Federal Statistical Office; Deutsche Bundesbank; calculations of DIW/ZEW

so clear-cut - appears if the shares of the foreign companies in the turnover and employees in Germany are considered (cf. *Table 4*). The share figures are relatively stable over the course of time - of the large sectors, only the chemical industry showed an increase in the importance of companies with foreign shares at the end of the 80s - and the differences between the R&D-intensive and non-R&D-intensive sectors was not so marked.

Table 4: Share of Turnover and Employees of the Foreign Enterprises in Germany 1980, 1990 and 1994

Branch	Turnover			Employees		
	1980[1]	1990[1]	1994	1980[1]	1990[1]	1994
	Share in %			Share in %		
Manufacturing industry	25,8	25,6	25,3	16,4	16,7	16,1
Non-R&D-intensive branches	27,9	25,0	23,9	14,2	13,4	12,6
R&D-intensive branches thereof:	23,3	26,2	26,7	18,5	19,4	19,3
Chemical industry	29,6	39,0	35,1	23,7	32,9	29,9
Mechanical engineering	16,7	16,4	18,3	14,2	13,9	15,2
Office machinery/ADP	72,2	73,7	85,7	49,2	53,6	52,4
Road vehicle constr.	19,8	22,0	24,3	19,0	17,2	18,1
Aerospace industry	14,4	15,9	22,2	11,1	5,8	9,2
Electrical engineering	22,8	23,2	22,5	18,3	16,8	16,6
Precision engineering, optics, clock-making	20,4	28,4	28,8	15,7	24,4	20,0

1) Former Federal Republic of Germany (pre 1990)
Sources: Federal Statistical Office; Deutsche Bundesbank; calculations by DIW/ZEW

These statistics alone should not tempt us into drawing conclusions already on the internationalisation of R&D in the sense of the third process of technological globalisation „International (global) Generation of Technologies". For this purpose the R&D activities of the companies abroad must be more closely examined.

The relative significance of foreign R&D activities varies considerably among the different countries. In *Table 5* some indicators are described. In Europe, 11.9% of the technological activities fall to foreign (i.e. non-European) large enterprises if the US patents which originated in Europe of these enterprises are taken as the criterion.

Table 5:   Indicators for the Relative Significance of Foreign Technological and R&D Activities in Selected Countries

Host country	Share of patents of foreign large enterprises 1990-94 in % (359 of the world's largest companies)[1]	Share of foreign R&D in % (1993 national surveys/ statistics)[2]	Share of foreign enterprises in manufacturing industry[3]
Japan	1.0	5.2	2.8 (1990)
Canada	----	40.6	49.0 (1989)
USA	4.2	15.3	14.8 (1992)
Europe	11.9	----	----
Belgium	49.1	----	----
Germany	9.3	15.9	13.8 (1992)
Finland	2.9	----	6.7 (1992)
France	11.4	15.2	26.9 (1991)
Great Britain	20.2	25.8	25.5 (1991)
Ireland	----	67.0	55.1 (1988)
Italy	9.6	----	22.3 (1988)
Netherlands	12.5	----	----
Austria	12.5	----	25.7 (1991)
Sweden	12.0	14,0	18.0 (1992)
Switzerland	5.4	----	----
Spain[4]	----	50,0	30 - 40

[1]   Source: Patel/Vega (1997), p. 7
[2]   Source: SV Scientific Statistics; US Department of Commerce; Beise/Belitz (1997); NIW, DIW, ISI, ZEW (1995); Barré (1996)
[3]   Source: OECD (1996), p. 36
[4]   Source:Buesa/Molero (1997)

For Japan this figure amounts to only 1% and for the USA 4.2%. Within Europe the shares of US patents which originated from the individual countries lie between 2.9% (Finland) and 49.1% (Belgium). In the midfield after Switzerland (5.4%) come Germany, France, Italy, Holland, Austria and Sweden with rates of around 10%, whereas the share for Great Britain is 20.2%. Statistics collected from individual countries, which are based on surveys in some cases, additionally show that the share of foreign R&D in the total R&D of the country in question is often greater than the share of the patents of foreign large enterprises originating from the country. Thus the share of foreign R&D in Japan amounts to 5.2%, in the USA, Germany and France between 15.2% and 15.9%, in Great Britain 25.8%, in Canada 40.6% and in Ireland 67%. At the same time, the share of foreign R&D is especially high in the countries in which foreign enterprises are strongly represented in manufacturing industry. On the basis of these discrepancies further investigations should be carried out to ascertain whether - and if necessary, for which reasons - the patent indicators underestimate the extent of the internationalisation of R&D and its increasing dynamics in the last few years.

Figure 1:   R&D Expenditure of Foreign Enterprises in the USA 1977 - 1994

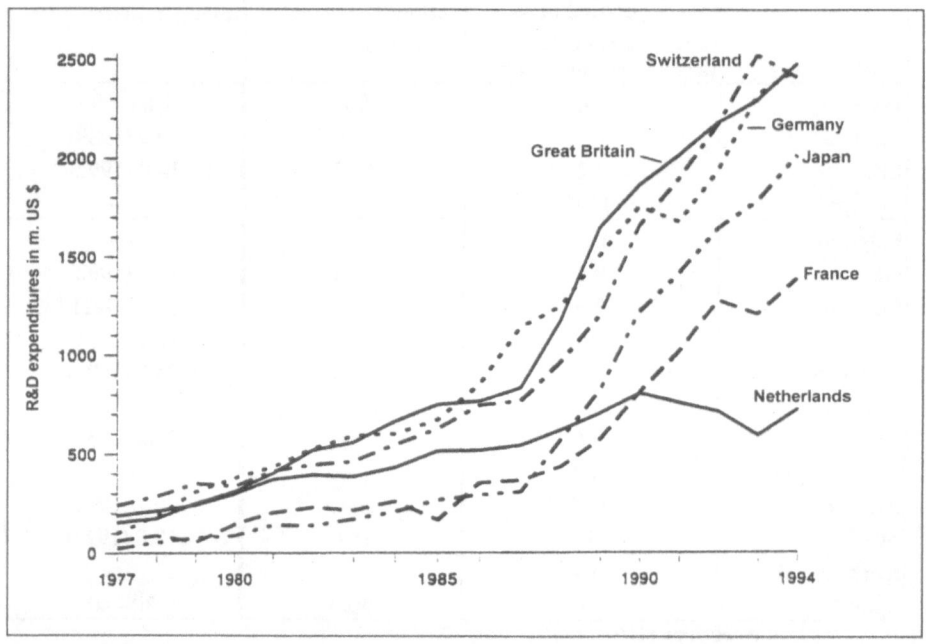

Source: US Department of Commerce, quoted by Beise/Belitz (1997)

In the regional distribution of the R&D activities abroad clear concentrations can be distinguished. Thus a special evaluation of the SV scientific statistics which is quoted in Beise/Belitz (1997), substantiates that the R&D activities of German enterprises are concentrated on Europe and the USA. Conversely, a large portion of the R&D expenditure of US American enterprises goes abroad to the European countries. According to the preliminary data published by the US Department of Commerce, the total US American R&D expenditure of the manufacturing industry abroad amounted to 10.147 bn US $ in 1994.[14] 25.9% thereof fell to Germany, 19.1% to Great Britain and 11.3% to France. Parallel to the focus on Europe, however, the US American R&D expenditure in Japan is increasing continuously: they had a share of 2.3% in 1982, in 1990 4.5% and in 1994 already 7.8%. The targeted destination of the Japanese enterprises performing R&D abroad is above all the USA. Europe takes second place.[15] The other Asian countries are also gaining in importance for Japan.

Even when the perspective is reversed, it is evident that the USA is a significant recipient country for foreign R&D expenditure. *Figure 1* illustrates that the industrialised states have continuously extended their R&D activities there. At the same time, with the exception of the Netherlands, in all the countries examined here the growth rate for the R&D expenditures increased in the second half of the 80s. In 1994 we find Germany, Great Britain and Switzerland in the lead, followed by Japan. The preliminary data of the US Department of Commerce for the R&D expenditures of foreign enterprises for the year 1995 indicate that the German enterprises have yet again significantly increased their expenditure and now spend approximately 3.9 bn US $ on R&D (1994: just 2.5 bn US $). They thus have clearly the greatest R&D potential in the USA, before Swiss and British enterprises. The great increase in R&D expenditure of German enterprises in the USA from 1994 to 1995 is primarily explained by the boom in company take-overs - especially by important acquisitions in the pharmaceutical industry - in which the R&D capacities also became part of the German enterprise.

A similar picture is found when the number of independent research centres belonging to foreign enterprises in the USA in the year 1994 are considered (cf. *Table 6*). Japan assumes the top position here, followed by Great Britain and Germany. If the technological foci of the foreign research centres are considered at the same time, it is seen that the enterprises conduct research in the USA mainly in those fields in which they have their technological strengths in their home country. So the centres belonging to Japanese and Korean companies concentrate on the fields of computers and electronics. The European enterprises place their main emphases in the sectors chemicals, pharmaceuticals, biotechnology and new materials. Furthermore, German and Dutch enterprises perform research above all in the electronics sector,

---

[14] The data of the US Department of Commerce and the regional distribution are taken from Beise/Belitz (1997), p. 45.

[15] Cf. Beise/Belitz (1997), p. 51.

Table 6: Independent Research Institutions of Foreign Enterprises in the US 1994 according to Branches

Branch	Japan	Great Britain	Germany	France	Switzer-land	Nether-lands	Korea	Sweden	Others	Total Industry
Pharmaceuticals, biotechnology	25	23	18	11	17	5	1	6	9	115
Chemicals, materials	24	19	28	17	10	4			8	110
TV, other electronics	33	10	9	4	5	4	4		3	72
Software	27	6	4	3		1	1		1	43
Computer	22		4			3	7		3	39
Semiconductors	19		3			2	10			34
Telecommuni-cation	15	2	4	2	1		1	2	3	30
Optoelectronics	11	2	3					1	3	20
Precision engineering, control engineering	1	23	3	6	6	3		1		43
Automobile construction	34	1	11	2	2		3	2		53
Metal-working industry	5	93	1	4	1			1	2	17
Mechanical engineering	7	4	2	3				6	5	27
Food and consumer goods	7	19	6	2	6	7		1	7	55

Source: Dalton/ Serapio (1995), quoted acc. to Reid/Schriesheim (1996). The columns contain multiple answers for institutions which conduct R&D in more than one branch.

Table 7:   Foreign Acquisitions by US High-tech Enterprises according to Branches and Countries, October 1988 till March 1993

	Materials	Aero-space	Chem-icals	Com-puter	Elec-tronics	Semicon-ductor equip-ment	Semicon-ductor	Tele-commu-nication	Biotech-nology	Others	Total
Japan	42	18	25	108	36	34	53	32	27	62	437
Great Britain	12	6	7	14	14	1	2	13	7	6	82
France	3	5	11	10	3	2	1	5	2	6	49
Germany	0	1	6	1	3	1	3	6	6	2	29
Canada	1	3	3	7	2	0	1	2	0	0	19
Switzerland	1	0	1	3	0	1	0	2	5	1	14
Taiwan	1	0	0	8	0	0	2	1	0	0	12
Australia	2	0	0	1	1	1	0	2	1	0	8
South Korea	0	0	0	4	1	0	2	0	0	0	7
Netherlands	1	0	2	2	1	0	0	0	0	1	7
Total	70	34	61	171	71	43	65	69	55	83	722

Source: Gaster/Prestowitz (1994), quoted according to Reid/Schriesheim (1996).

Table 8:   Patent and R&D Activities of Large Enterprises Abroad

	Geographical distribution of patents of large enterprises in the USA, acc. to countries of origin, 1990-94 in % (359 of the largest enterprises in the world) [1]						Percentage of R&D expen- diture abroad [2]
	Share in percent		Respective country abroad				Percentage share abroad
	Country of origin	Abroad	USA	Japan	Europe	Others	
Japan	98.0	2.0	1.4		0.5	0.1	2.5 (1993)
USA	92.3	7.7		1.0	5.2	1.5	13.0 (1994)
Europe	77.6	22.4	20.9	0.5		1.0	----
Austria	86.0	14.0	2.2	0.0	11.2	0.6	----
Belgium	33.2	66.8	12.5	0.0	52.5	1.8	----
Finland	64.9	35.1	8.0	0.0	26.9	0.3	----
France	67.3	32.7	18.0	0.4	13.2	1.0	----
Germany	79.6	20.4	14.1	0.6	5.1	0.7	18.0 (1995)
Italy	80.4	19.6	10.3	0.0	8.7	0.6	----
Nether-lands	39.7	60.3	30.4	0.9	28.4	0.6	----
Sweden	62.2	37.8	18.2	0.1	18.5	1.1	25.0 (1994)
Switzerland	44.7	55.3	29.0	0.7	24.8	0.7	50.0 (1992)
Great Britain	50.4	49.6	36.0	0.4	10.9	2.2	----
Total	87.4	12.6	5.6	0.5	5.6	0.9	----

[1] Source: Patel/Vega (1997), p. 9
[2] Source: SV-Wissenschaftsstatistik; OECD (1996); US Department of Commerce; Mataloni and Fadim-Nader (1996); Schweizerischer Handels- und Industrie-Verein (1994); Beise/Belitz (1997); NIW, DIW, ISI, ZEW (1995)

British companies in precision engineering, measurement and control technology and Swedish companies in mechanical engineering.

The distribution is almost identical with foreign acquisitions of US American high-tech enterprises in the time span from October 1988 and March 1993 (cf. *Table 7*). However, it is striking that the gaps between the individual countries with acquisitions in the USA are relatively large. While Japan made 437 takeovers in the period under consideration, France made only 82, Great Britain 49 and Germany only 29.

The increasing internationalisation of R&D provides in the final event only indications for the international (global) generation of technologies. A further convincing indicator is here - classified according to countries of origin - the geographical distribution of patent registrations in the USA of large enterprises that operate abroad (cf. *Table 8*). Accordingly the international generation of technologies is mainly brought about by European enterprises. Japanese enterprises take almost no part in this process and US American companies practically none. The European enterprises on the other hand distribute their activities in the USA and European countries.[16] Great differences can be determined between the various European countries, however.

The international technological collaboration to develop know-how and innovations with partners in more than one country, who retain their institutional identity, can also be represented by indicators. However, only relatively old data are available on technological co-operations between enterprises. So Hagedoorn/Schakenraad (1990), based on the Cati-Merit database, find a massive growth of new technological co-operation agreements for the 80s in the relatively new technology fields of biotechnology, information technology and new materials.

Table 9:   International Distribution of Technological Co-operation Agreements in the Fields of Biotechnology, Information Technologies and New Materials (Number, Percentages in Brackets), 1980 - 1989

	Biotechnology	Information technologies	New Materials	Total
Western Europe	223 (18,4)	509 (18,7)	118 (17,2)	850 (18,4)
Western Europe-USA	245 (20,2)	599 (22,0)	133 (19,3)	977 (231,2)
Western Europe-Japan	38 (3,1)	177 (6,5)	49 (7,1)	264 (5,7)
USA	428 (35,3)	707 (26,0)	139 (20,2)	1274 (27,6)
USA-Japan	155 (12,8)	406 (14,9)	94 (13,7)	655 (14,2)
Japan	58 (4,8)	95 (3,5)	88 (12,8)	241 (5,2)
Others	66 (5,4)	225 (8,3)	67 (9,7)	358 (7,8)
Total	1213 (100)	2718 (100)	688 (100)	4619 (100)

Source: Hagedoorn and Schakenraad (1990)

If the geographical distribution of the international technological agreements is taken into consideration (see *Table 9*), then it becomes clear that the greatest accu-

---

16 The conclusion reached by Archibuge/Michie (1995), p. 134, that the international generation of technologies is mainly an internal European affair cannot not be confirmed by the more recent data.

mulation of such joint ventures takes place in the USA. In 2.906 technological co-operation agreements (that is, 63% of all the agreements recorded in the database), at least one company located in the USA participates. In addition, European

Figure 2:   Number of Publications Produced with German Co-authorship
                 1980 - 1996

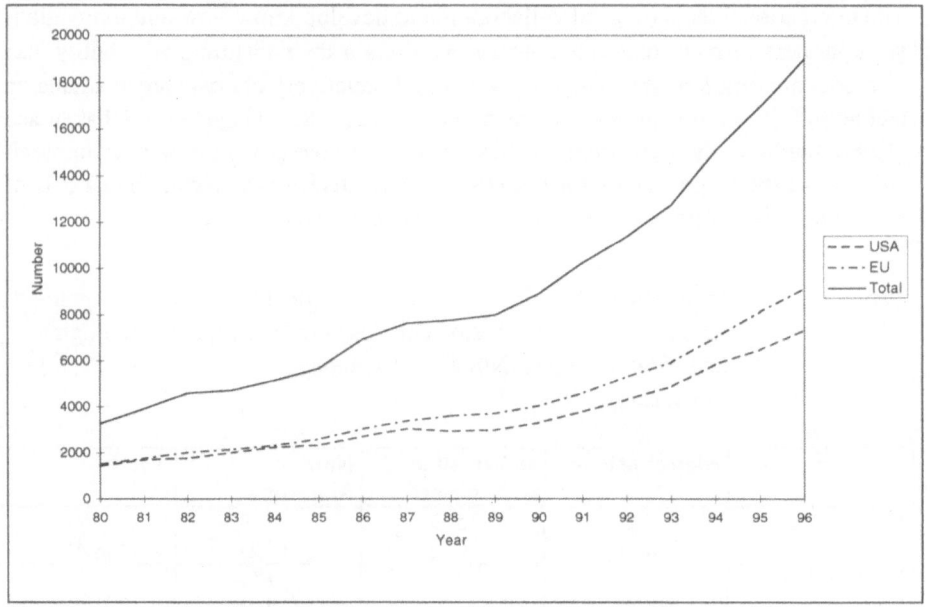

Source: Science Citation Index

enterprises co-operate more with US American enterprises (total share 21.2%) than with other European enterprises (total share 18.4%). Japanese companies prefer also technological co-operations with enterprises which have a location in the USA. Co-operations with enterprises in locations outside the Triad play only a subordinate role, with a share of 7.8%.

If the actors are scientists from universities and public R&D institutions, then the degree of international technological collaboration can be approximated by the number of publications written by international co-authors. In *Figure 2* the number of publications co-authored with German participation as recorded in the Science Citation Index is given. In total, the number of such publications has risen from 3.281 in 1980 to 19.023 in 1996. In 1996 7.341 publications appeared with American co-authors and 9.125 publications with co-authors from other European countries.

The Science Citation Index tends to exaggerate the chronological development, as new journals are constantly being included in the records, so that an increase in the absolute numbers is not necessarily accompanied by an increase in the actual publications.

Figure 3:   Share of Co-publications of German Authors with Foreigners in the Total Publications of German Authors 1980 - 1996

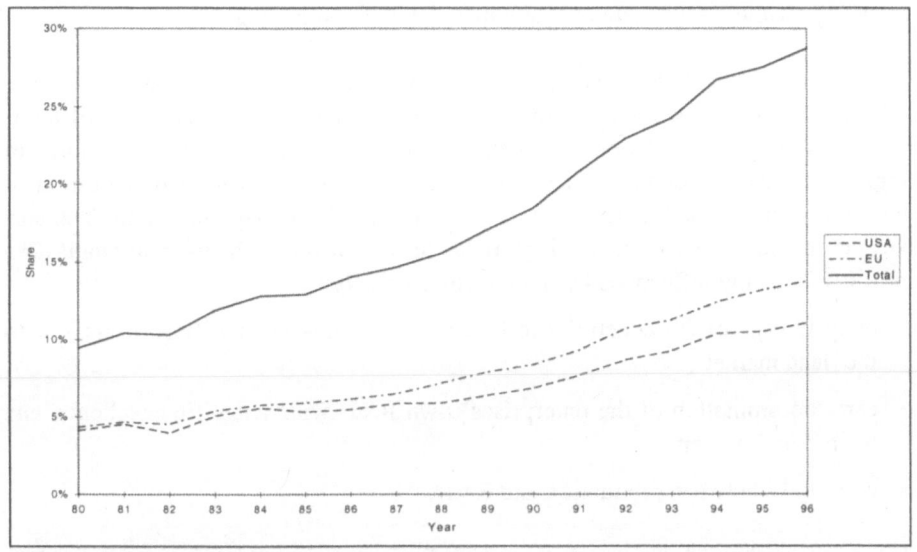

Source: Science Citation Index, own calculations

It is therefore practical to weight the absolute figures with the relevant number of the total publications in which German authors were involved or which were written by German authors alone. These percentages of publications written in international co-authorship with German participation are presented in *Figure 3*.

When considered relatively, the marked increase of publications written by international co-authors with German participation is also evident. The share of these publications in the total publications by German authors was 9% in 1980, and increased steadily to 29% in 1996. Until the middle of the second half of the 80s, German authors published almost equally with US American and European co-authors, but now the co-authors from other European countries have gained in importance (1996: US American co-authors 11% and co-authors from other European countries 14%).

2.2   Internationalisation of Industrial R&D
in Selected Technology Fields

The models of internationalisation of industrial R&D in three key technologies - pharmaceuticals, semiconductor technology and telecommunications - have been analysed in-depth in a recent study carried out by ISI, DIW and ZEW (1997) for the BMBF. A number of results emerged from this study which advanced and complemented previous analyses and are summarised in the following.

The internationalisation of enterprises is more advanced in some branches than in others. Differences between sectors regarding the degree of liberalisation of international trade, the regulation of streams of direct investments, specific features of regional demand, economies of scale in production and the internationalisation of technological knowledge, result in different levels of internationalisation. The surveys in the three selected technology fields have shown that the internationalisation of R&D is mainly influenced by three factors, namely:

- early linkage of R&D activity to leading, innovative clients ('lead users') or to the 'lead market',

- early co-ordination of the enterprises' own R&D with scientific excellence and the research system,

- close links between production and R&D.

Our analysis showed that internationally active enterprises think in terms of value-added chains and process chains. Consequently, the criteria for selecting a location for R&D include not only factors of supply, such as a well-developed research infrastructure, but also demand factors, which are increasingly playing a more important part in the decisions of enterprises. Only by linking various value-added chains can (relatively) non-transferable 'performance alliances' be created, establishing Germany internationally in selected fields as a location for competence centres which it would be difficult to transfer, or duplicate, elsewhere.

The importance of lead markets in anchoring existing industrial R&D activities and attracting new activities has increased. The market's function as a 'lead market' is decisive for innovations which only fully mature when they come into close contact with demanding, innovative customers. In fields of technology that are strongly science-based, it is the results of scientific research that constitute a driving force in the internationalisation of innovation processes. In both cases, regional proximity to external partners such as customers, competitors and scientific institutions is an advantage. If there is a close interlinking of production and R&D activities, internationalisation of R&D follows internationalisation of production. The internationalisation of production is then the main driving force behind the internationalisation of R&D.

One central finding of this survey is that the determinants of internationalisation in the three fields of technology considered are different (cf. Table 10). The dynamics of innovation in product development in semiconductor and in software in tele-communication technology is largely driven by lead markets.

Table 10: Determinants of the Internationalisation of R&D in Selected Fields of Technology

Importance of R&D link to	Pharmaceuticals		Semiconductor technology		Telecommunications technology	
	Pre-clinical	Clinical research	Process technology	Product developm.	Hardware	Software
Lead market	low	very high	low	very high	low	very high
Science/ re-search system	very high	high	high	low	high	low
Production	low	low	high	low	high	low

In process technology in semiconductor technology and in hardware in telecom-munications the linkage of production with R&D is also a significant factor. In the pharmaceutical industry a clear distinction has to be made between pre-clinical and clinical research: the innovation dynamics in pre-clinical research are driven by scientific excellence, whereas in clinical research it is the lead market that is the driving force. The link of R&D to production is very loose in this case.The follow-ing section describes these differences for the three fields of technology.

Pharmaceuticals

Pharmaceuticals, as engaged in by research-performing producers of medical drugs, can also be characterised as a science-based sector. The following features are in-dicative of its science-based nature: (1) in pharmaceuticals, performing one's own R&D is decisively important for innovativeness and competitiveness, (2) universi-ties and other research institutions constitute an important source of information for innovations, and (3) innovations are strongly based on the results of basic research. In addition, the pharmaceutical industry can be described as having a level of inter-nationalisation of R&D which is high compared to other sectors. This is true for the shares of patents abroad (i.e. from outside the enterprise's home country), for R&D expenditure abroad, and for the various instruments used to acquire technology such as take-overs, joint ventures and the acquisition of qualified personnel - in all of which activities abroad plays a very marked role. Particularly in the subsector of biotechnology, the generation of knowledge is highly internationalised.

It emerges from the study that a clear distinction has to be made between the determinants in the choice of an R&D location for pre-clinical and clinical research (cf. Table 10). In the pre-clinical phase (exploratory research and chemical development) the generating of new medicines is driven by scientific excellence and the most recent research results. Innovation dynamics at the interface of pharmaceuticals/biotechnology are influenced by excellent universities, research institutions and biotechnology start-up firms. It is important here that all the players should command highly developed co-operation and management skills and that there should be no barriers between the various institutions, since otherwise no transfer of scientific knowledge or technology can take place. The excellence of regional research in the form of innovative, research-performing start-up firms, universities, clinics and other research establishments offers far stronger incentives to attract the research activities of multinational enterprises to a region than is the case in either of the other two fields of technology investigated.

In clinical research (including pharmaceutical development, clinical studies) a close link with the lead market is the decisive factor when selecting a location. The quantitative and qualitative significance of the market for the product concerned is important, as well as satisfactory co-operation with approval authorities, a high degree of transparency in approval criteria and an appropriate infrastructure in the clinics which can ensure professionalism in conducting the clinical studies.

A close link between R&D activities and production is not a determinant for research-performing producers of medical drugs when selecting a location for R&D; production is organised relatively independently from R&D.

Highly internationalised pharmaceutical enterprises generally pursue a concept of "Triadisation" of their locations, which also includes R&D activities (cf. Figure 4). Following the phases of the innovation process a decision must again be made between exploratory research and chemical/pharmaceutical development (including clinical research). Exploratory research is mainly organised on a 'matrix' principle: here competencies are distributed by therapeutic areas among a few (2-5) excellent research centres. To do this, transdisciplinary areas such as biotechnology, combinatory chemistry and other methods are defined, which are controlled centrally and world-wide for the corporation from one location.

Product development is no longer controlled from the home country but globally. There are two organisational alternatives here (cf. Figure 4): firstly, a global product development centre may be set up at one location, assuming world-wide responsibility for the strategic orientation and global control of all development projects. Secondly, 2-3 product development centres may be established in various locations world-wide, each having global responsibility for product development in specific therapeutic areas. In both alternatives, clinical studies in Phase II are carried out *in situ* in the most important countries or markets at an *operative* level.

Figure 4:   „Stylised facts": Organisation of Research and Development in a Highly Internationalised Pharmaceutical Enterprise

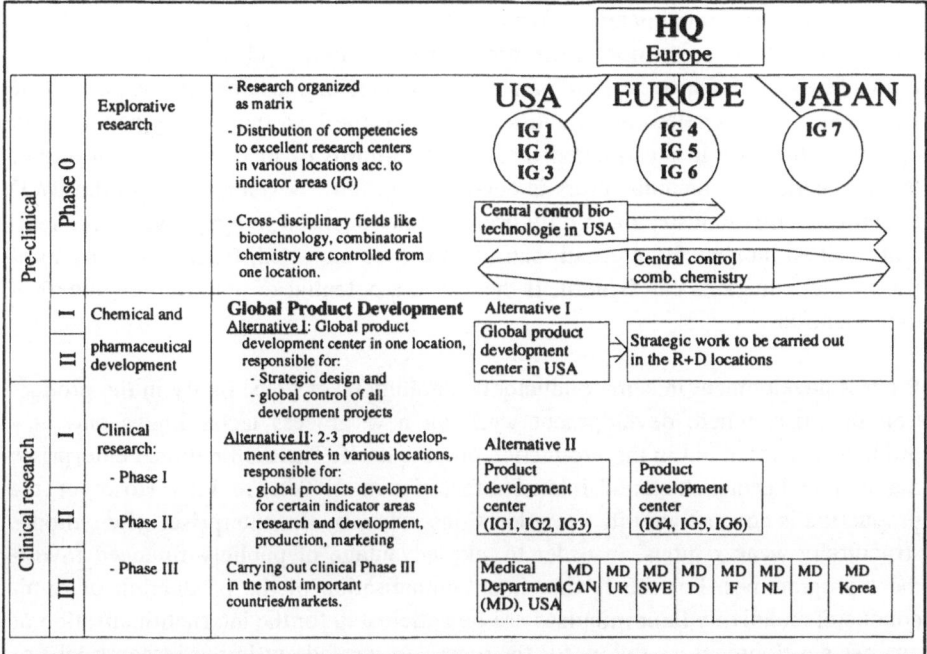

Source: ISI, DIW and ZEW (1997)

From the viewpoint of a country or region and from a policy viewpoint, the establishing of a research centre or global product development centre is highly significant, because it is here that the relatively non-transferable 'performance alliances' are created. These alliances become much less easily transferable if research and product development centres both coincide within the same region or location. This is the case, for instance, with foreign pharmaceutical firms in the regional innovation clusters in the USA.

Semiconductor Technology

In semiconductor technology, the scientific linkage is slight (cf. Table 10). For the most part, it emerged primarily in the development labs of the large enterprises, in the US also in small, newly founded companies which grew with the market for semiconductor components. With the diffusion of the technology into different areas of application, the importance of collaboration between semiconductor manufacturers and customers which have system knowledge also grew. These innovative enterprises create markets for new products, in which the close co-operation with the semiconductor manufacturers, and close proximity to them, is a prerequisite for the competitiveness of the partners. Thus the proximity to the lead markets is the decisive driving factor for the internationalisation of the development of application-specific products (embedded chips) in semiconductor technology. In the past years, lead markets for semiconductor components have emerged in Europe for the automotive, telecommunication and Smart Card areas. Above all semiconductor enterprises which up to now were not internationalised and had concentrated on only a few application areas (e.g. personal computers or consumer electronics), have had to establish development facilities abroad in the lead markets as part of their diversification. As it is estimated that over two-thirds of the R&D expenditure of the semiconductor manufacturers is devoted to product development and the R&D activities of this industry have only been internationalised to a relatively small extent, a sharp increase in internationalisation can be expected, from which the locations in lead markets can benefit, if they have a potential of qualified engineers at their disposal.

Process development in semiconductor technology takes place partly in the production facilities, where development work on new process technologies and new products is carried out in the production process itself. The multinational enterprises maintain and establish capital-intensive factories in all Triad regions. However, the production is concentrated in a few locations, whereby the enterprises often choose structurally weak regions, in order to take advantage of publicly financed investment programmes. For the further internationalisation of the production of semiconductors relatively faint impulses can be anticipated for the internationalisation of process development, as this is for the most part carried out in the research labs of the enterprises at their home location.

Telecommunications Technology

In telecommunications technology the most influential factor for internationalisation is the proximity to the lead markets, in which the new technologies which can catch on in other countries, or even world-wide, are first utilised (see Table 10). The standardisation process, without which a broadly based communication network is not possible, plays a crucial, amplifying role. The science base of telecommunications can only be observed in new technological subsectors. As a rule, the large enterprises have sufficient internal competencies at their disposal as a result of their many years of research, to develop telecommunication equipment further. The maturity of this technology is apparent in the smaller number of technical possibilities available. Impulses for innovations come from the application front, i.e. from the users. Public basic research is of subordinate importance in this case. The regional proximity to the research facilities however cannot be transformed into a decisive lead in development for the enterprises, if the link-up to the lead markets is not given.

The coupling of production and research is only observed to a small extent in telecommunications. Above all, due to the large part played by software, which already accounts for two-thirds of the value added, the separation of R&D activities and the production of equipment (hardware) is much more easily possible than in technologies which depend to a greater extent on the testing of prototypes and the interaction between production and product development. As the development of software is usually independent of the hardware, the internationalisation of software development takes place independently of the internationalisation of production. The R&D activities are in many instances more internationally distributed than the production. The hardware development (product development) is still concentrated in the manufacturing locations of multinational enterprises.

2.3   Motives and Consequences of the Globalisation of R&D

The empirical findings show that not only the degree of internationalisation of the exploitation of new technologies, but also of research, product development and innovation itself has increased continuously in the 80s and 90s.[17] This globalisation of research, innovation and technology markets is comparatively far advanced in branches and product segments with a high generation of knowledge and a strong country-specific differentiation of products and research systems. Until now, the pace in the globalisation process has been set by certain segments of the chemical/pharmaceutical industry (particularly agricultural chemicals, pharmaceutics, biotechnology) and the information technology industry (semiconductors, EDP, telecommunication, consumer electronics). There is still some 'catching up' with the trend towards internationalisation of R&D to be done in branches where production and assembly still constitute a substantial part of value added, such as the automobile industry and the construction of industrial plants and machinery, but as the 1990s progress these, too, are being swept along by the current of globalisation. Enterprises that are very far advanced in globalisation in specific branches are already showing counter-tendencies towards 'de-globalisation', as growing complexity makes efficient steering more and more difficult.

The most important motives for the continuation of the trend towards globalisation in R&D and innovation activities are:

(1)   Access to leading research results and talents;
(2)   Presence on-the-spot, learning in lead markets and adaptation to sophisticated customer needs;
(3)   Initiation and strengthening of R&D at locations where the effects of greatest usefulness can be expected and the highest cash flow is generated;
(4)   Monitoring and taking advantage of regulatory frame conditions and standardisations;

---

17 Cf. Gerybadze, Meyer-Krahmer and Reger 1997.

(5)   Support of production and sales on-the-spot by local R&D capacities.

Thus the primary motive and aim of the internationalisation of R&D is not - as it has been in the past - the simultaneous maintaining of several globally 'dislocated' R&D units, but the globalisation of learning processes along the whole of the value-added chain (research, development, production, marketing/sale, service relations, embedding in supply and logistic networks). The decisive parameter for the intensity of transnational learning and innovation processes is the proportion of value added within the corporation constituted by the generation of knowledge.

Many leading enterprises - and especially German corporations - are planning to expand their R&D capacities world-wide in the medium term, but generally not in their countries of origin. In view of a *latent growth potential of world-wide research capacities*, for which the locations have not yet been decided, national technology policy will have to orient itself more strongly towards the strategies of enterprises from other countries. This would offer chances for Germany to attract foreign enterprises to locate R&D here, and to build up centres of competence.

2.3.1   Establishing World-wide Competence Centres for R&D

Whereas the 1980s were a period during which the internationalisation of R&D was associated with decentralisation and the 'dislocation' of activities, the 1990s are characterised by a continuing trend towards internationalisation, accompanied by concentration, focusing and strategic emphasis. International enterprises that are leading performers of R&D are pursuing the strategy of a presence with R&D and product development at precisely those locations where there are the best conditions world-wide for innovation and the generation of knowledge in their product segment or field of technology. They are no longer satisfied with locations which 'just about keep up' with the global technology race; they deliberately seek out the unique centres of excellence.

Although the majority of large international enterprises performing R&D are still following the strategy of keeping the competence base for their core technologies in their country of origin, processes of re-thinking are in progress. The dynamics of change in this context are dependent on global technology strategy on the one hand and, on the other, on the size and the resource base of the country of origin. The largest Swiss chemical firms internationalised their R&D earlier, and to a much greater extent than, for instance, the German ones. Thus within a branch or product segment, a broad distinction can be made between two patterns: in corporations with a strong research and market base in their country of origin, units abroad mostly continue to have only scanning and exploration functions as well as tasks of applications development (this is true particularly of enterprises originating in Japan, in the USA, and in Germany with the exception of chemicals/pharmaceutics).

Compared with these, corporations with a less developed research and market base in their country of origin have come to occupy a 'vanguard' role in globalisation. In corporations with their headquarters in Sweden, the Netherlands or Switzerland, and also in some individual enterprises from the large industrialised countries, R&D activities are increasingly being shifted to centres of excellence abroad, and the idea of concentrating 'core technologies' in centres of competence abroad is also definitely being considered.

Even in large international corporations, this world-wide focusing strategy and formation of centres is associated with considerable adaptation measures in organisation and management. The absorptive capabilities of an organisation, which enable it to draw sustained benefit from centres of excellence abroad, depend on whether the enterprise itself has concentrated enough competence on the spot, and whether it provides support from headquarters in the form of resources and decision-making competence. Despite their growing importance in terms of R&D expenditure, R&D units abroad in many enterprises still do not receive sufficiently strong strategic support and are sometimes inadequately co-ordinated. In the 1980s the linking of internationalisation with decentralisation led to duplication of tasks, to R&D units lacking the 'critical mass' of resources and capacities, and to disputes about competency. From these experiences, transnationally oriented enterprises are now going over to consistent, cross-corporate technology management. This generally also implies that the core activities of their R&D are concentrated as far as possible in one place and assigned as clearly as possible to responsible groups and locations.

Table 11:   Orientation of R&D according to Degree of Innovation

R&D	Incremental innovation	Radical innovation
global	Development of equal parts	Centres of excellence and lead markets
local	Adaptations to local/ national conditions	Dissemination of start-ups

Source: Gerybadze, Meyer-Krahmer and Reger (1997, 201)

The outcome is that this development leads to the fixing of just one centre as a 'leading house' for one specific product group or technology within a corporation, as far as possible. In view of this, the competition between innovation systems will increase. For allocation decisions in R&D, this change of direction implies that excellence of a national research system, although a necessary prerequisite for these decisions, is not in itself a sufficient condition. Conditions that have to be satisfied include particularly the presence of lead markets, in the case of radical innovations (cf. Table 11). With incremental innovations, it is mainly a case of building up local R&D capacities for the support of production and sales.

2.3.2   Formation of High Performance Units and 'Clusters'

The relationship between production and R&D locations has become much looser. Competitiveness of production locations obviously tends to be based on the 'cluster formation' of regional production structures and supply networks, nearness to big markets and minimisation of factor costs, rather than the presence of lead researchers, whose knowledge, wherever it has been produced, can now be used worldwide. On the other hand the concentration of industrial research in a very few 'centres of excellence' in the world gives a small number of science centres the opportunity to offer themselves as attractive locations. The implications for technology and innovation policy are:

(1)   Combining the attractions of the market location with those of the production location and the R&D location, and exploiting the advantages of these links, works very effectively in enhancing attractiveness for the allocation decisions of internationally oriented enterprises.

(2)   However, making use of the attractiveness of these three types of 'Standort' in combination, and realising sustained synergies, can probably only succeed under very special conditions. The decisive factor is the type of driving force behind the synergy relations. The generation of cash flow is one powerful driving force. Production, as a cash flow generator, thus brings R&D in its wake, as for instance in the pharmaceutics industry and the automobile industry. Due to their cash flow, large and interesting markets still have good chances of becoming production locations as well as important R&D locations.

From time to time, contrary developments can also be observed: In areas where the dynamics of technological change are weak and/or where there are no substantial synergies between product- and production-related knowledge, R&D locations and production locations may well become dissociated. On the other hand, for certain types of strategies - particularly in highly dynamic fields - the close linkage of both locations is important. Under certain conditions, all three functions (market, production, R&D) may even coincide in one location. In the latter case, both from the viewpoint of the investing enterprise and of the location being invested in, only those projects and development strategies can have a sustained and really positive impact, in which functioning high performance units are established along the whole length of the value chain. Under these conditions, R&D laboratories are set up primarily where the best conditions are to be found world-wide, both for research and also for the transfer of its results. These R&D units are part of a functioning cycle in the host country, and at the same time are embedded in a highly effective network of transnational learning.

2.3.3 The Variety of Co-ordination Mechanisms

Following an initial phase of over-enthusiastic decentralisation of R&D in the 1980s, growing problems of co-ordination led to disillusion and the increasing formation of centres in a global context. At present, many multinational enterprises are experimenting with various mechanisms for steering and integration, with the aim of creating synergies world-wide and avoiding the duplication of tasks. It can be regarded as certain that, to co-ordinate global R&D activities, an intelligent set of mechanisms is needed which must be combined as effectively as possible. Whereas a number of Japanese enterprises investigated place the emphasis on personal contacts, informal communication and socialisation, combined with a centrally dominated decision-making process, the Western European enterprises in surveys mainly rely on contract research for the divisions and daughter companies as a coordination mechanism. [18] Particularly in the German enterprises, the importance of informal instruments and the formation of a corporate culture is often underestimated.

Particular importance attaches to the use of 'hybrid' co-ordination mechanisms (such as multifunctional, interdisciplinary projects, strategic projects, technology platforms, core programmes and core projects). The novel aspects of these coordination mechanisms are that they cut across - or overlay - organisational and hierarchical structures, and that they are often used for the simultaneous coordination of several different aspects - for instance, co-ordination of R&D strategy with business strategies, integration of the business functions of R&D, production and marketing, as well as ensuring synergies between various areas of technology. In the enterprises investigated, R&D units abroad are often not involved in strategic projects; so far this is only the case in a few, truly transnational enterprises.

Manifold requirements for co-ordination also exist in public research systems. In this context, it can be observed in several countries that the development of new, flexible types of co-ordination mechanisms is not nearly so advanced in the public research systems as it is in the enterprises investigated. Several approaches, some of them newly developed and some already tried and tested, can also be transferred in adapted form to meet new networking needs in the public research system. This particularly applies to hybrid and informal co-ordination instruments, which can be used to form networks between various different levels and types of actors.

---

[18] Cf. Reger (1997).

2.3.4   Management of Corporate Research and Future Business

In the transition from the first generation of R&D management (dominance of central research) to the second generation (divisionalisation, subordination of research to divisional interests) most large international enterprises substantially weakened their basic research in the course of the 1980s. At the beginning of the 1990s, the third generation of R&D management tried to achieve a kind of synthesis (simultaneity and equilibrium of group development and basic research, formation of portfolios). Empirical investigations show, however, that third generation management of R&D is causing problems in all the enterprises to a greater or lesser extent, and that up to now various models have been experimented with, all of which have to be regarded as 'second bests'.

Japanese corporations are particularly consistent in their way of opening up promising future areas that require many years of preliminary research. A new research laboratory with a clear mission is set up, well-equipped in terms of staff and financial resources. As soon as a topic shows promise of becoming marketable, the laboratory is affiliated to an existing division; the new technology is used for the expansion of existing fields of business. Alternatively, the laboratory forms the nucleus for a new division, if the enterprise has not previously been active in the relevant market. Several good examples of this establishing of R&D laboratories abroad, and the subsequent founding of 'spin-offs', can be found in Canon, Mitsubishi Electric, Sharp and Matsushita Electric.

In any case, it can be stated that the enterprises are frequently trying to establish a balance between central research and development in divisions or business groups; no 'best practice' for this has been found so far. In large Japanese enterprises, excellent use is made of basic research abroad as an instrument for opening up promising fields of business in the long term. This example not only demonstrates the importance of global 'technology sourcing' but also shows that judicious linking and embedding into the research systems of other countries is a necessary practice. Thus enterprises and research institutions, in their efforts to achieve a stronger international presence in this way, will necessarily enter the orbit of national technology policy.

As a general result of this situation, the premise of national science and technology policy, encountered in many countries; that the main benefit from the public allocation of resources in this policy area flows into the national economy is progressively dissolving. Not only the know-how produced in the national innovation system, but also other public investments, for instance in training and education, are increasingly being swept into the stream of the international exchange of knowledge. This development enlarges the focus of policy: it is not simply the appropriation of nationally generated knowledge that is involved, but the strengthening of a generally beneficial, interactive transnational exchange of knowledge. It is possibly as important to absorb knowledge that has generated world-wide, as it is to support the production of knowledge in one's own country.

3.   Consequences for National Innovation Policy

It would be a fallacy to assume that national innovation policies will lose in significance as a result of globalisation. The international exploitation of innovations requires that national governments create conditions under which new technologies can be exploited within the respective countries. International technological co-operations are based on the national technological capabilities and possibilities available to the co-operation partners.[19] The international generation of innovations requires efficient national innovation policies, which meet the new challenges. This view is supported by new results obtained in innovation research, which show that in the face of increasing international mobility on the part of enterprises and technologies, as well as the growing harmonisation of vital pre-conditions (infrastructure, human capital), efficient national innovation systems will have an increasingly significant role to play in the organisation and promotion of innovations.[20]

3.1   Education, Research and Technology Policy Measures

The discussion on regional locations is often suggestive of the notion of defending a national fortress (the stronghold of the nation as a location for industry, or other national domains such as science and culture) while extending, in a one-way process, the possibilities of its industrial subjects to conduct their business in other countries. However, globalisation is forcing the course of events in another direction: it implies a mutual opening-up and 'penetrability' of legal and economic frontiers, of science and research systems, mobility of people, cultures, organisation and management systems. A pro-active national technology policy will therefore also open up to other countries' enterprises and research establishments.

A considerable number of foreign enterprises are actively performing research and development in Germany - some of them in their own R&D laboratories. The idea of the national science system being opened up to foreign enterprises, or foreign research establishments being set up is frequently associated with the fear that antennae are simply being installed to 'siphon off' nationally accumulated knowledge. These enterprises and establishments are considered with reserve, since there are

---

[19] Cf. Archibugi/Michie (1995), p. 134.

[20] Cf. Porter (1990), Lundvall (1992) and Nelson (1993).

fears of a one-sided drain of science and technology to headquarters abroad. It is feared that the 'knowledge and technology drain' may take place without there being any positive impacts for the national innovation system, and that in the long term it will serve only to enhance the innovativeness and competitiveness of foreign rivals. However, it is not so much the geographical situation of the parent enterprise that is decisive for the impacts, as what type of R&D activities, what production capacities and services locate in the host country (e.g. autonomous research versus local antenna, highly-skilled manufacturing as opposed to the 'extended workbench').

There is still a lack of clarity regarding the impacts of foreign R&D units on the national or regional location. However, the decisive factor is probably not the ownership situation so much as the willingness of foreign enterprises to establish the whole value chain, including research and development. A few US studies have shown that the R&D performed within a national economy is increasingly exploited world-wide, so that the idea that national technology policy primarily causes positive effects in its own country is no longer applicable. A stronger inclusion of foreign enterprises into national technology policy is thus inevitable in the end, and the issue at stake is to shape this process as usefully as possible for the home country. Japan, for instance, supports the presence of industrial R&D in its own country. 'Useful' in this context implies *the generating of as many spill-over effects as possible within the country*. The involvement of Sony, for example, in regional DAB (Digital Audio Broadcasting) pilot projects in Germany is leading to a build-up of high-grade R&D capacities, and also possibly production capacities.

With the help of a matrix, the R&D activities at the location can be subjected to a first evaluation (cf. Fig. 5). If both the autonomy and the competence of the local R&D are low, it can be described as a 'local antenna'. Local antennae monitor the newest technological and market trends and transfer information to the corporation's country of origin; such transfer is *one-way* (Case 1). If autonomy is low but competence is high, the R&D management is characterised by centralisation of the decision-making process (Case 2). Although R&D activities are carried out autonomously, the appreciable domestic spill-over effects will probably be only moderate, due to the centralised decision-making. If autonomy from headquarter is high but competence is low, knowledge tends to be exploited on-the-spot (Case 3). This type of R&D is usually associated with production-supportive technology centres and the exploiting of local market chances. If the competence and freedom of decision of the local R&D unit are both high, the unit is a centre of R&D competence which also contributes to integrated transnational R&D activities. In this case (Case 4), it may definitely prove useful to include it more strongly in national technology policy. With regard to cases (2) and (3), the advantages and disadvantages more or less balance out; in case (3), at least, gains in competence can lead to positive development into a real, leading R&D centre within the corporation, which is also beneficial for the location.

67

Some countries, such as Great Britain, Canada and Singapore for instance, pursue a deliberate policy of attracting foreign R&D. Depending on the technological specialisation of national industry, different patterns of 'location policy' can be observed: the US, as a world leader in many research areas, behaves as a 'bastion' in order to avoid too great a science and technology drain. Japan, by contrast, is (still) pursuing the course of a 'claim', still trying to isolate itself by 'soft' access barriers. Other countries, such as Singapore, pursue a strategic location policy, aiming to attract foreign R&D by focusing on specific fields and building up centres of competence.

Figure 5:   Matrix for Evaluation of Local R&D Units

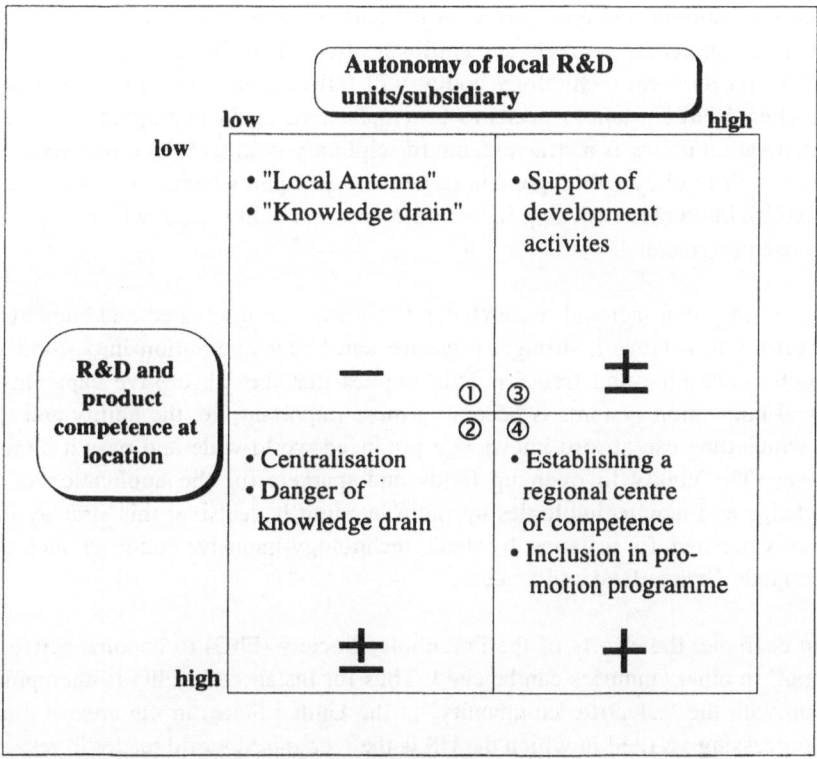

Source: Gerybadze, Meyer-Krahmer and Reger (1997, 213)

In the context of regular reporting to the Federal Ministry of Economics on the structure of industry, two reports were submitted recently. One study by the HWWA-Institut für Wirtschaftsforschung (1995) comes to the conclusion that the globalisation of German industry (primarily with regard to production) implies a growing importance for industrial policy. With the internationalisation of production increasing, improving the quality of locations would mainly mean improving the qualification and flexibilisation of the workforce, promoting investment and accelerating public decision-making. According to this report, the financial support of domestic enterprises (i.e. enterprises with their headquarters in Germany), including the public promotion of technology, are increasingly missing the mark, since it is not certain whether these measures will generate income in the national or regional locations.

An investigation by Gerybadze, Meyer-Krahmer and Reger (1997) comes to similar conclusions (e. g. on the subsiding of R&D) with regard to this question of whether traditional technology policy is 'on target' or not. However, it is precisely this circumstance which leads the authors to plead the case for a re-formulated concept of technology promotion, namely: both to support national research institutions and enterprises on their path towards globalisation and, at the same time, to gain foreign research institutions and enterprises for the national innovation system and, in both cases, to attain synergy effects and spillover effects beneficial to the location. The fact that, on its own, technology policy will fall into an 'inadequacy trap' under these altered circumstances needs to be emphasised again and again. Technology and innovation policy is a strategic, interdisciplinary task, and the effectiveness and success of this policy will depend in large measure upon whether it proves possible to establish internal networking in this field between policy areas which have previously been fragmented.

It is necessary that national research establishments be motivated and supported in their efforts to achieve a stronger presence and better integration into world-wide research networking and transfer. This implies that the 'absorptive capability' of national innovation systems is becoming more important, i.e. the ability and speed with which they can absorb knowledge produced world-wide and pass it on to enterprises. The ability to open up fields and markets for the application of new knowledge and new technologies by rapid learning is decisive; this strategy is deliberately pursued, for instance, by small, technology-intensive countries such as the Netherlands, Switzerland and Sweden.

As an example, the efforts of the Fraunhofer Society (FhG) to become active 'on-the-spot' in other countries can be cited. Thus for instance the FhG is attempting to link up with the 'scientific community' in the United States in the area of graphic data processing - a field in which the US is the recognised world leader in research - where it is trying to achieve the position of a seriously regarded partner. By contrast, in the area of production technology, particularly lasers, the FhG is rather pursuing the aim of presence in a market which is of increasing global importance for FhG services. The FhG has now also intensified its activities in South East Asia, where it is treading new ground by taking on the role of an international 'broker' between technology supply and demand.

'Learning' from other research and innovation systems relates to the behaviour of enterprises, particularly the reduction of 'home-grown' deficits such as inadequate linkage of R&D to the market, too much concentration on technology and insufficient optimisation of technology use, organisation and qualification - i.e. conscious efforts to strengthen transfer competence. A forward-looking attitude with regard to the acceptance of technology belongs equally to learning. For example, in the case of foods produced by genetic engineering there is an urgent need for enterprises to enter into an intensive dialogue with consumer associations. This is happening in the Netherlands and Denmark, but not in Germany. The commonly made assertion of a dominant and widespread attitude of technology rejection in this country obscures the fact that science, state and industry have often omitted to explain the impacts of their actions at an early stage or discuss them sufficiently early with society. In other innovation systems, acceptance of technology is better prepared by bringing in the affected parties at an early stage. The list of such examples of success can be extended indefinitely. Early anticipation and active influencing of market acceptance enhance the success chances of innovations.

Learning relates to structural changes. Thus it emerged from a comparative analysis by ISI on transfer systems in the US and Germany that, despite considerable differences in the two national innovation systems, they have many similarities, and that technology transfer instruments are therefore transposable from one system to the other.[21] From a German viewpoint, for instance, more active marketing of patents at universities, improving the framework conditions for venture capital, and intensifying co-operation between national large-scale laboratories and industry along the lines of the American CRADA model, are of interest. On the other hand, however, regarding stronger mobilisation of industrial sponsorship capital for universities, the limits of German 'sponsorship culture' would probably be reached very quickly. From an American perspective, the models of German 'An-Institute' (extramural-type institutes close to universities with various forms of organisation and status), the Fraunhofer institutes and industrial joint research can give impulses for the improvement of the US's own system. In both systems, public institutions fulfil a central function. Following the trend towards globalisation, they will have to open up more than they have done so far to the participation of foreign research institutions in national programmes.

It is decisive for the German innovation system that it should enable efficient transfer and rapid learning to take place through intelligent interlinkage, in order to pursue the strategies of the rapid second innovator, become seriously regarded as an international 'player' and intelligently transfer structures, processes and framework conditions that foster innovations. This recognition brings with it a number of specific implications for technology policy, of which some examples are cited here. Some of these have already been elements of S&T policy for many years now, whereas others set new policy accents:

---

[21] Cf. Abramson et al. (1997).

(1)   Supporting the international activities of national public R&D institutions and enterprises by:
- Establishing international training/education and research programmes,
- Fostering the international mobility of students and scientists (in Germany, personnel exchanges have been found to be in the 'under five percent' zone) as well as encouraging researchers and students from abroad to come to Germany,
- Supporting the presence of domestic research institutions in other countries (joint ventures with other research establishments, research teams or institutes on a temporary basis),
- Supporting enterprises in their efforts towards a stronger global presence in R&D (including the acceptance of this strategy),
- Building up technological competence and positioning as an international 'player' to be taken seriously in areas which have not been among the country's classical strengths so far.

(2)   Supporting the location of foreign R&D establishments.

(3)   Two-way incentives such as
- Promoting transnational projects (e.g. further developments of the EUREKA type),
- Supporting the 'brokerage function' of public research institutions, to support the international exchange of technology supply and demand,
- Creating framework conditions and structures conducive to innovation on national and regional levels,
- Monitoring innovation-friendly structures in other countries and making use of this experience for national policy.

It is necessary, especially for the consequences mentioned under point (1), that the German universities are reformed. Research and teaching must become increasingly the suppliers of ideas and motors for innovation for our society. Globalisation increases the competitive pressures also on the university system, so that the quantitative survival of the education system must go hand-in-hand with qualitative improvements.

Measures to improve Germany as a location for higher education, which would also lead to an increase in the number of foreign students, are investments for the future in the coming scientific and economic co-operation with other parts of the world. Discussion partners and decision-makers abroad, who have known Germany since their university days, are 'door openers' for industry. In this connection the package deal planned jointly by the Federal and Länder governments to create stronger incentives for foreign students to obtain further qualifications at German universities is greatly to be welcomed.

The consequences catalogued under point (3) recommend that technology promotion should further encourage the formation of innovative networks between enterprises and research institutions as a reaction to globalisation. Not only leading-edge research in isolated fields, but the tapping of broad innovation potentials linked with this research by enterprises contribute to the international attractiveness of the German innovation system. Not all enterprises - and this applies specially to SMEs - are in the position to develop and implement their own globalisation strategies. They are therefore all the more dependent on being included in regional and European networks. Joint projects undertaken by science and industry (referred to by the Federal government as 'lead projects') on a national and European level can help to build up a critical mass and create promising networks of competence. These projects should bundle together demanding terms of reference with a perspective for concrete application and join together various disciplines and possible application fields. They should be suggested and worked out by the partners in a 'bottom-up' fashion, as is already the case in the EUREKA Programme.

The European project „Prometheus", for example, provided significant impulses for the meantime high level of German industry in the development of traffic control technology. Joint projects in microelectronics have contributed to the fact that the industry in Europe has re-established a position in the production of microelectronic components. The target of a new promotion project on human genome research is to establish leading-edge research in one of the most dynamic science fields worldwide. At the same time, academic and industrial research are being linked up more closely than in the past. A high degree of attraction and a maelstrom effect of German human genome research is aimed for in international comparison. In teleservices the idea is that using the most up-to-date information and communications technology, manufacturers can carry out maintenance work, diagnoses and even repairs world-wide on machines that are located at a great distance from the production plant, without a service engineer having to be on site. The solution of such fundamental technological tasks, just as the building up of a effectual infrastructure, can open up new competitive prospects.

The development of regional economies of scope as a basis for establishing regional networks and competence centres should be supported by appropriate incentives on the part of innovation policy. The BioRegio competition was tested as an instrument to this end. In the 80s, great scientific potential in biotechnology was built up by research policy. Among the 50 research institutions most frequently quoted in this technology field, 8 German ones are presently to be found. However, this potential has not been sufficiently linked up to the strong chemical and pharmaceutical industries. In the competition, 16 regions competed among themselves for the best preconditions to turn biotechnological knowledge into products, processes and services. A great number of initiatives were started decentrally to achieve the best possible networking of knowledge. Biotechnology in Germany today is unmistakably at a time of new departures: German companies which years ago relocated their research and manufacturing facilities abroad, are investing in Germany again today, and researchers are returning. Dynamic regions, such as Munich or the Rhine-Neckar triangle, radiate new competence in this technology to the whole world and demonstrate that technology locations can carve out a new profile for themselves.

In this connection, the role of small and young enterprises is of particular significance for the innovative strength of an economy. The dynamism of these companies, especially in the new technologies, has a substantial influence on the location quality. Fortunately, in the last few years the venture capital market for „early phase" financing has developed positively in Germany, and has led to numerous new business start-ups, e.g. in the field of biotechnology. This was also encouraged by a number of public special programmes, for the normal promotion process for setting up new businesses is not always sufficient. Very high investment costs and risks on the one hand are often faced with problems of access to equity capital and outside capital on the other. The tapping of new financing possibilities for innovative enterprises therefore has a high priority. If someone can demonstrate good ideas, competence and the willingness to shoulder entrepreneurial risks in new fields responsibly, then the access to the necessary capital in Germany should not fail. If private investors have been increasingly encouraged in the past by promotion programmes to invest in innovative companies, it is now vital that the frame conditions for the venture capital market be improved, for the long term.

But also in the more traditional branches of the economy small and medium-sized enterprises are being confronted with the impacts of globalisation and must meet the challenges of international competition. Thus a survey e.g. carried out recently by DIHT (German Industrial and Trade Association) in the foreign trade chambers in 72 countries showed that medium-sized enterprises also follow in the footsteps of the large companies. In most cases the supplier firms see a great potential in this step to maintain their previous business relations to large enterprises.

3.2   The Significance of "Lead Markets"

Analysis of the innovation activity of transnational enterprises shows that they are increasingly thinking in terms of integrated process chains, and are not primarily transferring their value added to places which provide the best conditions for research only.[22] The demand side obviously plays a more important role in R&D allocation decisions than do supply factors. From a macroeconomic viewpoint, the central question is rather: 'Where will income be generated, where will benefits be felt and where will new resources be created?' than: 'Where will costs be created and where will existing resources be consumed?' In their transnational investment

---

[22] Cf. Gerybadze, Meyer-Krahmer and Reger (1997, especially chap. 3).

activities, enterprises are acting according to the following decision patterns: Where are the attractive, future-oriented markets in which users can be learned from, and which generate a sufficiently high return-on-investment for costly product development? Where can these markets be best served by highly developed production, logistic and supply structures? Where would it therefore be worthwhile to build up value added in *one* place? In what countries do attractive markets, highly developed production structures and excellent research conditions coincide, so that innovative core activities can be concentrated there?

In view of the strategic decision processes in transnational enterprises, the determinants and motives we have identified raise the following questions for national technology policy:

(1) In what end-user markets is the country regarded as a trend-setter, both in Europe and internationally?
(2) In what regions are production structures and supply networks so highly developed that high value-added can be secured for the innovation system as a location in the long term?
(3) What areas of the regional research and technology system are at a leading level world-wide and can also induce effects of strengthening national/regional lead markets and production structures?
(4) Where is influence being exerted (through participation in research and standardisation alliances, or in complex learning processes taking place in a national and/or regional context) on 'dominant technological designs' for innovations, which will subsequently bring lead advantages in the global innovation competition?
(5) What is the relative strategic importance of the country as a market, and as a production location, from the viewpoint of enterprises world-wide?

By creating effective links with these fields of competence and building up 'forward-backward linkages', it may prove possible to create high performance units with low transferability which are unique by world standards. Only by combining excellence in research with highly developed European lead markets, or by combining research with highly developed production structures, can the national innovation system position itself as a location for core competencies that are not readily internationally transferable.

What are the characteristics of lead markets? They match one or more of the following criteria:

(1) a demand situation characterised by high income elasticity and low price elasticity or a high per capita income,
(2) a demand with high quality requirements, great readiness to adopt innovations, curiosity concerning innovations and a high acceptance of technology,

(3)   good frame conditions for rapid learning processes by suppliers,

(4)   authorisation standards that are 'trailblazing' for permit authorisation in other countries (e.g. pharmaceutics in the US),

(5)   a functioning system of exploratory marketing ('lead user' principles),

(6)   specific, problem-driven pressure to innovate,

(7)   open, innovation-oriented regulation.

The attractiveness of the national innovation system from this perspective is determined not so much by comparative, static competition factors such as costs and wages, as by its 'dynamic efficiency'[23]. This is largely dependent on the extent of social and organisational intelligence in the finding and acceptance of new structures and markets. Will complex system innovations (such as road pricing, product/service packages, closed-cycle economic concepts, new applications for information technology) be elaborated e.g. in Germany which will be used world-wide? Offensive learning through numerous field trials and pilot schemes for the finding of technical, economic, legislative and social solutions is important. Learning processes of this kind often take years. The innovation system that first succeeds in mastering these complex solutions gives participating enterprises competitive advantages, and appears more attractive to foreign investors.

Globalisation is forcing national (and European) technology policy to re-focus technology promotion, orienting it towards the initiation of complex innovations that reach far into economic, legislative, social and societal domains. Here, too, it is the pace of learning and the mastery of new solutions that count. Not only leading edge research, but the opening-up of new (lead) markets by anticipatory, future-oriented pilot projects is decisive for the international attractiveness of the national innovation system ('keeping ahead in the learning race'). The target group for technology policy has altered: research-driven enterprises are engaging in a change of strategy and are giving more consideration to the conditions of lead markets and production networks. Technology policy will scarcely be able to avoid following this change.

For this reason, successful R&D locations have particularly good chances of inducing economically positive impacts, e.g. on employment, if they coincide with production and market locations. Technology policy *on its own* cannot constitute a policy strategy promising success, this is an inherent dilemma. Thus the results of our study underline the necessity - much called for, but not as yet fulfilled - for better networking between different areas of policy. Insofar as they influence science and technology, these include - to name but a few examples: (1) fiscal framework conditions for the formation of venture capital, (2) the phasing-out of subsidies that preserve the *status quo*, (3) regulation and approval procedures that relate to spe-

---

23 Economic theory differentiates between static efficiency - relating to one point in time - and dynamic efficiency - relating to a long term development. It is quite possible for static and dynamic efficiency to conflict with one another.

75

cific results and not to specific techniques, (4) departmental policies such as transport, health, planning, environment, and economic policy, (5) an active competition policy and (6) increased flexibilisation of civil service law in order to enhance the flexibility of institutionally-supported research establishments. To match the ever-increasing international demand for complex innovative and high performance units/networks, the lateral structures essential for their formation must once more be called for in policy.

References

Abramson, H. N./Encarnacão, J./Reid, P. P./Schmoch, U. (Eds.) (1997): *Technology Transfer Systems in the United States and Germany*, Washington, D.C.

Aharoni, Y. (1971): On the definition of the multinational corporation, *Quarterly Review of Economics and Business*, 11, pp. 27 - 37.

Archibugi, D./Michie, J. (1995): The globalisation of technology: a new taxonomy, *Cambridge Journal of Economics*, 19, pp. 121 - 140.

Barré, R. (1996): Relationships between multinational firms' technology strategies and national innovation systems: a model and an empirical analysis, in OECD (1996a): *Innovation, Patents and Technological Strategies*, Paris.

Beise, M./Belitz, H. (1997): *Internationalisierung von Forschung und Entwicklung in multinationalen Unternehmen*. Materialien zur Berichterstattung zur technologischen Leistungsfähigkeit Deutschlands 1996. Berlin, Mannheim.

Buesa, M./Molero, J. (1997): *La estructura industrial de España*. Madrid.

Dunning, J. H. (1977): Trade, location of economic activity and the MNE: A search for an eclectic approach, in Ohlin, B./Hesselborn, P. O./Wijkman, P. M. (Eds.): *The International Allocation of Economic Activity*, London.

Dunning, J. H. (1979): Explaining changing patterns of international production: In defence of the eclectic theory, *Oxford Bulletin of Economics and Statistics*, 41, pp. 269 - 295.

Dunning, J. H. (1981), Explaining the international direct investment position of countries: Towards a dynamic or developmental approach, *Weltwirtschaftliches Archiv*, 117, pp. 30 - 64.

Europäisches Patentamt (1996): *Jahresbericht 1996*, München.

Gerybadze, A./Meyer-Krahmer, F./Reger, G. (1997): *Globales Management von Forschung und Innovation*, Stuttgart.

Hagedoorn, J./Schakenraad, J. (1990): Inter-firm partnerships and co-operative strategies in core technologies, in Freeman, C./Soete, L. (Eds.): *New Explorations in the Economics of Technical Change*, London/New York, pp. 3 - 37.

HWWA-Institut für Wirtschaftsforschung (1995): *Grenzüberschreitende Produktion und Strukturwandel - Globalisierung der deutschen Wirtschaft*, Forschungsauftrag des Bundesministeriums für Wirtschaft, Hamburg.

Hymer, S. H. (1960): *The International Operations of National Firms: A Study of Direct Foreign Investment*, Cambridge (Mass.), 1976 veröffentlicht.

ISI, DIW, ZEW (1997): *Internationalisierung industrieller F&E in ausgewählten Technikfeldern*, unveröffentlichter Entwurf des Endberichts an das Bundesministerium für Bildung, Wissenschaft, Forschung und Technologie (BMBF); vorgelegt durch das Fraunhofer-Institut für Systemtechnik und Innovationsforschung (ISI), Deutsche Institut für Wirtschaftsforschung (DIW), Zentrum für Europäische Wirtschaftsforschung (ZEW), Karlsruhe, Berlin, Mannheim.

Jungmittag, A. (1996): *Langfristige Zusammenhänge und kurzfristige Dynamiken zwischen Direktinvestitionen und Exporten* - Eine mehrstufige Modellierung dynamischer simultaner Mehrgleichungsmodelle bei kointegrierten Zeitreihen, Berlin.

Lilienthal, D. (1960): Management of the multinational corporation, in Anshen, M./Bach, B. L. (Eds.): *Management and Corporations 1985*, New York, pp. 119 - 158.

Lundvall, B. A. (Ed.) (1992): *National Systems of Innovation*, London.

Mataloni, R. J./Fadim-Nader, M. (1996): Operations of US multinational companies: Preliminary results from the 1994 benchmark survey, *Survey of Current Business*, December, 11-37.

Münt, G. (1996): *Dynamik von Innovationen und Außenhandel*, Entwicklung technischer und wirtschaftlicher Spezialisierungsmuster, Heidelberg u. a.

Nelson, R. (Ed.) (1993): *National Systems of Innovation*, New York.

NIW, DIW, ISI, ZEW (1995): *Zur technologischen Leistungsfähigkeit Deutschlands*. Erweiterte Berichterstattung 1995. Zusammenfassender Endbericht an das Bundesministerium für Bildung, Wissenschaft, Forschung und Technologie (BMBF); vorgelegt durch das Niedersächsische Institut für Wirtschaftsforschung (NIW), Deutsche Institut für Wirtschaftsforschung (DIW), Fraunhofer-Institut für Systemtechnik und Innovationsforschung (ISI), Zentrum für Europäische Wirtschaftsforschung (ZEW), Hannover, Berlin, Karlsruhe, Mannheim.

OECD (1996): *Globalisation of Industry. Overview and Sector Reports,* Paris

Patel, P. and Vega, M. (1997): *Technology Strategies of Large European Firms,* Interim Report for 'Strategic Analysis for European S&T Policy Intelligence' Project funded by EC Targeted Socio-Economic Research Programme. Brighton.

Porter, M. (1990): *The Competitive Advantage of Nations*, London.

Reger, G. (1997): *Koordination und strategisches Management internationaler Innovationsprozesse*, Heidelberg et al.

Reid, P. P./Schriesheim, A. (1996): *Foreign Participation in U.S. Research and Development - Asset or Liability*, Washington, D.C.

Schmoch, U. (1996): International patenting strategies of multinational concerns: The example of telecommunications manufacturers, in OECD (1996a): *Innovation, Patents and Technological Strategies*, Paris

Schweizerischer Handels- und Industrie-Verein/Bundesamt für Statistik (1994): Forschung und Entwicklung in der schweizerischen Privatwirtschaft 1992, Zürich.

UNCTAD (1996): *World Investment Report 1996*: Investment, trade and international policy arrangements, United Nations, New York, Genf.

Emerging Global Economic Trends and Issues[1]

Ulrich Hiemenz, Olivier Bouin, David O'Connor,
Dominique van der Mensbrugghe

1.   Introduction

The first three post-war decades saw an historically unprecedented rise in OECD living standards, resulting in no small measure from closer integration of Member countries' economies. Growth in the OECD area has slowed in the last two decades and its revival will depend on further liberalisation of their economies, including through domestic regulatory reform and other structural adjustments. High on the agenda is an increase of the flexibility of OECD markets - to facilitate adjustment to changing word market conditions without jeopardising macro-economic stability - and continued international liberalisation of trade and capital flows - to expand and deepen the integration of world markets. To realise the growth potential associated with the process of globalisation will depend as much on economic policy reforms in the OECD region as on complementary reforms in developing and transition economies outside the OECD area (henceforth referred to as non-member countries). In recent years, a growing number of non-member countries have come to recognise the potential offered by market-oriented reforms - involving stronger links with the global economy - to promote rapid economic development and improved living standards for their people. As their reforms bear fruit in faster growth, their importance as trade and investment partners of the OECD countries will grow steadily in the coming decades.

Against this background, the OECD Secretariat has undertaken to assess the policy reform agenda facing member and non-member countries alike and to explore - in a quantitative way - the implications of continued globalisation for the world economy over the next 25 years (OECD, 1997). Key questions were: What are the economic benefits to be produced by globalisation? Is it worth the effort? Is continued globalisation feasible and sustainable or are there going to be bottlenecks in food or energy supply? What about social cohesion in the face of accelerated structural adjustments? And, will poor countries be marginalised by international economic integration?

---

[1] This paper represents a summary of the OECD study "The World in 2020: Towards a New Global Age", published in November 1997. The study is based on contributions from all substantive OECD Directorates, the International Energy Agency and the OECD Development Centre which also had the overall drafting responsibility.

To be sure, the OECD Secretariat has not attempted to forecast future develop-ments. A brief look back in history tells that real world events are always subject to political or economic shocks which cannot be predicted. Cases in point are the two oil shocks or the demise of socialism in Eastern Europe which have changed the course of global economic development in many ways. Instead of projections, the OECD Secretariat has chosen to develop alternative long-term scenarios for the world economy, based on alternative assumptions of how successful the non-member countries - but also OECD countries - are in continuing economic reforms and putting in place the necessary policies and institutions to sustain high growth over the next 25 years. Such scenarios allow to evaluate potential outcomes of pol-icy reform efforts and to analyse side effects of continued globalisation.

Section 2 of this paper summarises major driving forces of globalisation while Sec-tion 3 provides an overview of the economic policy challenges facing Member and non-member countries. Depending on the pace with which policy reform is pro-gressing, the world economy could have very different features as shown by the scenarios presented in Section 4. Section 5 evaluates some of the implications as-sociated with these scenarios, and Section 6 gives a brief summary of major results.

2.   Major Driving Forces of Globalisation

Spurred by strong innovation and fast diffusion of technological change as well as by institutional and policy reforms - at both national and international levels - globalisation has accelerated in recent years. These forces are likely to provide a continued impetus to closer integration of economies in the years to come.

• Low-cost computing, communications and transport have been instrumental not only in linking non-member countries more closely to the outside world, but also in facilitating domestic commerce. Satellite, microwave, cellular radio, and fibre optic transmission are only a few of the communications technologies that have contributed to major cost reductions and accelerated expansion of communica-tions networks, including in low-income countries. Low-cost computers have made fairly advanced information processing power available even to small businesses in member and non-member countries alike, while the Internet has given those businesses a degree of access to world markets that would have been inconceivable even a few years ago. Still, the potential benefits from advances in global communications and computer networking are far from having been fully exploited and the effects of these technologies on global production and trade patterns have only begun to be felt.

• Growing economic links between OECD and non-member economies are an increasingly important feature of globalisation, owing much to a sea-change in development strategy and economic policy in the non-member countries in the direction of sound macroeconomic policies, greater economic openness, a larger role for the private sector, and increased reliance on market competition. The East Asian growth experience has provided much of the inspiration for such policies. In recent years, a growing number of non-member countries have become sizeable exporters of manufactures to OECD countries, and there is now a flourishing two-way manufactures trade. A growing share of OECD's trade is with non-member countries, and the same is true of private capital flows. China, India, Russia, Brazil, Indonesia - referred to in this study as the Big Five - are already large economies and their weight as importers, producers for world markets, hosts to foreign investment, and foreign investors is bound to increase.

International institutional and policy developments also point towards closer global economic integration. Tremendous progress has been made in the last few decades towards a freeing of trade and capital flows: quantitative trade restrictions have been dismantled in many countries, tariffs have been substantially reduced, exchange controls relaxed and investment regimes liberalised. Regional initiatives aimed at trade liberalisation and broader economic integration have proliferated. At the same time, multilateral economic institutions have seen their membership enlarged and their powers enhanced. Completion of the Tokyo and Uruguay Rounds of trade negotiation has led to broad-based tariff reductions and the easing of some of the important non-tariff barriers. The establishment of the WTO in 1995 (supplanting the GATT) has greatly strengthened the permanent institutional mechanisms for the discussion of trade issues and the resolution of disputes. Its membership has grown from 88 in 1985 to 131 at present and both China and Russia have defined their policies and programmes with a view to gaining admission.

3.   The Policy Reform Agenda for a New Global Age

3.1   The Unfinished Agenda of Trade and
Capital Market Liberalisation

Even with the considerable progress towards an open international trading and investment regime, there remains much unfinished business. On the trade agenda, with the very substantial reduction of border barriers once Uruguay Round commitments are fully implemented, emphasis will likely shift to "behind-the-border" measures that have traditionally been the domain of competition policy - e.g., regulatory reform, privatisation and other measures to enhance domestic competition. Trade policy and domestic policies to spur competition and ensure a "level playing field" have a high complementarity that is increasingly important in shaping approaches to international negotiation. For countries at comparable levels of development and per capita income, mutual recognition agreements (MRA) are one approach that has been used to address the problems arising from variable environmental, product and other standards and regulations. The experience with regional trade agreements suggests the possibility of generic blueprints detailing practical steps for reaching MRAs. Reliance on international standards, where they exist, when designing new domestic standards and regulations is another approach. In general, it is advisable to define a framework of minimum criteria to ensure that domestic regulation is consistent with promoting greater competition and enhancing market access.

On the investment agenda, main tasks are to liberalise capital flows further, to strengthen prudential supervision of capital markets and to level the playing field for investors irrespective of nationality. Although restrictions on capital movement have been almost entirely eliminated in the OECD area, important regulatory barriers to portfolio investment by certain types of institutional investors - private pension funds and life insurance companies in particular - remain and should be phased out. This will require that a suitable balance be struck in OECD countries among three objectives: (i) to give institutional investors greater options to raise returns; (ii) to minimise risks of large capital losses, notably to pensioners; and (iii) to discourage excessive risk-taking by fund managers (the moral hazard problem). The challenge to policy makers in the non-OECD countries is to draw up a comprehensive programme for the removal of capital controls, the liberalisation of market access and domestic regulatory reform, without compromising financial stability. The transition to liberalised capital flows necessitates a careful examination of the macroeconomic policy framework, the state of the financial system, and the ability of product and labour markets to adjust to domestic and external financial shocks.

Further efforts at liberalisation of investment regimes should promote non-discrimination between domestic and foreign investors. The Multilateral Agreement on Investment (MAI) currently being negotiated among OECD Members can be expected to attract accession by non-members interested in affirming their commitment to high standards of treatment of international investment. Other aspects of international investment requiring attention include tax neutrality for cross-border investments; protection of the corporate tax base in the presence of high capital mobility; and incentive competition to attract foreign investment.

3.2   Policy Challenges for OECD Countries

While OECD economies have enjoyed significant real income improvements over the past 25 years, and governments have achieved significant progress in managing inflation, trend productivity growth has been declining and structural adjustment has made uneven progress. It has become increasingly clear that key macroeconomic problems affecting many OECD countries - notably high unemployment and slow growth - are largely of a structural nature. Despite some progress and limited success in some countries, the policy responses to these macro-structural challenges have to date been generally insufficient. In the future, demographic trends could reduce growth prospects as would failure to achieve fiscal consolidation and further competition-enhancing structural reforms - notably product market deregulation, social welfare and labour market reforms. If growth is to advance, OECD Member countries will have to attend to their own domestic reform agendas, and put in place the policies that will make their economies and societies more adaptable to the forces of change while maintaining social cohesion.

- *Enhancing competition and innovation.* In the advanced OECD societies, a flow of cost-reducing, productivity-enhancing and market-oriented innovations is crucial for sustained economic growth. Innovation is also a leading element in OECD's comparative advantage as a global competitor and trader. Experience shows that competition is critical in stimulating such innovation. While most technology will continue to be generated in the OECD area, the challenge will be to enhance OECD competitiveness in the face of new pressures from new, stronger competitors, many of which will be seeking to move upmarket in terms of the technology embodied in the products and services. By streamlining regulations, de-restricting entry into sectors once reserved as "natural monopolies" and privatising state-owned enterprises, governments can help domestic industry become more cost-efficient and innovative. Recent OECD work suggests that regulatory reforms could yield significant productivity gains in certain sectors and boost the level of GDP over the long run by up to one per cent in the United States, 4.5 per cent in the UK, and 6 per cent in Japan, Germany and France.

- *Reforming the social welfare system.* The rather comprehensive social welfare systems in OECD countries were mostly designed in the 25-year period of high growth following World War II. Since the early 1970s, falling growth rates and rising unemployment have put social security systems under heavy pressure. Thus, the urgent policy challenge is to design a set of labour market, education and social policies which facilitate the capacity of individuals to adapt to structural change and to assist those who fall behind - all in the context of severely constrained budgets.

- *Coping with population ageing.* The ageing of OECD Member countries' populations will accelerate in coming decades, with important implications for savings rates, government budgets and debt positions, interest rates and growth rates. The ratio of pension fund contributors to beneficiaries has already declined in many countries, resulting in higher tax and/or debt burdens. A key element of the policy response is likely to be a higher effective retirement age, which would not only reduce pension outlays but broaden the contribution base. Closer targeting of state pension benefits on poorer retirees could also yield substantial system savings in some countries. A shift towards funded pension schemes (from the current pay-as-you-go systems prevalent in many countries) should pay off in the future, since real interest rates are likely to remain rather high and investments in fast-growing non-member countries will provide high returns.

3.3 Sustaining Reforms in Non-member Economies

A brief glance at the record of the last twenty-five years shows that only a handful of countries, almost all in East and Southeast Asia, have sustained high GDP growth rates - averaging 5 per cent per annum or more - throughout this period. The dynamic Asian economies have demonstrated the effectiveness of sound macroeconomic policies, outward orientation, and investments in human capital for sustaining high growth over long periods of time. Their linkage-oriented policies have not only been good for growth, but also for poverty alleviation. In the set of poor countries, mainly in Asia, which over the last three decades oriented their economies towards dynamic participation in world trade, investment and technology flows, a profound transformation of living standards has been achieved. No case is more dramatic than that of China where the share of the population below the poverty line more than halved since 1978 to reach 27 per cent in 1994. In sharp contrast, the poverty reduction record has been disappointing and even negative in those countries, whether relatively advanced or relatively poor, that remained too long with economic policies and structures unfriendly to market forces and international trade and investment.

A growing number of non-member economies - including the Big Five (Brazil, China, India, Indonesia and Russia) - have embarked on a path of far-reaching reforms design to put in place a growth-inducing policy package, namely, low inflation, "realistic" exchange and interest rates, low fiscal deficits, the strengthening of competitive markets, increased outward orientation, and reduction of government's ownership in the economy. Many reforming economies have made great progress in the battle against macroeconomic instability. Trade and investment have been substantially liberalised, though far more so in some countries than in others. Governments must now turn their attention to the most challenging parts of the reform agenda - the completion of domestic deregulation of both product and factor markets, privatisation, deepening of the financial sector, strengthening of the framework of law, property rights and institutions to foster domestic entrepreneur-ship and innovation, and improvement in the quality of public institutions and services, not least education. In some respects, notably the emphasis on microeconomic and structural issues, the challenges facing non-member countries increasingly resemble those facing OECD countries.

In various specifics, however, important differences remain. The educational attainment of the population in most non-member countries still lags quite far behind the OECD countries and in some there are still wide gender gaps. Higher investments in education will be necessary over extended periods if the education gap, including the gender gap, is to be narrowed significantly. Moreover, many non-member countries have grossly inadequate infrastructure, which adds significantly to the costs of doing business and discourages both foreign and local investment. Continued institutional and policy reform to foster market competition and more efficient resource allocation would positively impact growth in the non-member economies not only through higher productivity growth but also through faster capital accumulation and higher investments in human capital.

4.   The World Economy in 2020

4.1   Assumptions

Based on the above-mentioned reform agenda of OECD and non-member countries, two sets of growth assumptions for fifteen regions of the world have been devised (for the regional concordance, see Annex 1). They form the basis of alternative scenarios - "high growth" and "low growth" - that reflect differential progress with national and international policy reforms. Under "high growth", governments would make substantial further progress with domestic policy reforms as well as with global trade and investment liberalisation (achieving free trade by 2020); in the low growth scenario, they would not. Globalisation and its associated growth dividend could turn out to be even greater than assumed under the "high growth" scenario. By the same token, the "low growth" scenario - corresponding to "business as usual" - is by no means a worst case, since it envisages neither major economic disruptions resulting from natural or political events nor a return to wholesale protectionism. The implications of the two scenarios for future per capita incomes, trade and production patterns, employment and earnings, food and energy markets, and the global environment have been explored in a consistent modelling frame-work. The scenario results are indicative, providing insights into the various driving forces that might influence global outcomes, and flagging problems to be anticipated and opportunities that could be exploited.

Based on best estimates of growth impacts from the reforms associated with the high growth scenario, OECD GDP is assumed to expand over the next 25 years at nearly the same rate as in the past 25 years (*i.e.*, on average 2.8 per cent per annum, figure 1 and Annex 2). The negative impact on economic growth from ageing - which in the extreme case of Japan could slow GDP growth by as much as 0.6 per cent per annum over the next 25 years - would be more than offset by strong productivity growth. Four main factors could boost productivity growth: regulatory and other competition-enhancing reforms; trade and investment liberalisation; technological progress; and the upgrading of human capital. On the other hand, a combination of weak policy reforms and a less open global economy would result in a 0.7 percentage point reduction in growth over the 25-year period.

Figure 1:   GDP Growth Rates (percentage points of GDP growth)

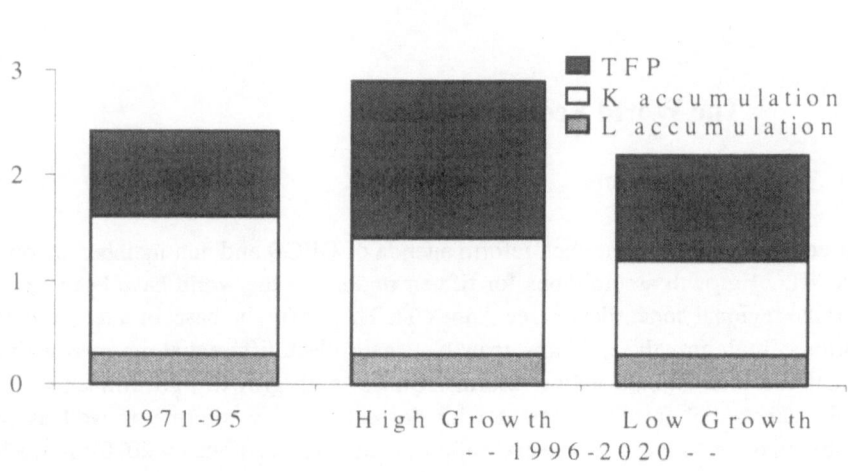

OECD

NMEs

Source: OECD estimates

In non-member economies, underlying growth potentials are significantly higher and sound policies should provide an additional growth impulse. In the "high growth" scenario, GDP growth among non-member economies would average 6.4 per cent per annum (2.5 percentage points more than in the past 25 years), though with considerable regional variation. Productivity is expected to grow at twice the

OECD rate, reflecting largely technological catch-up and intersectoral reallocation of labour from low-productivity to high-productivity sectors. Considering the negative productivity growth in much of sub-Saharan Africa since the early 1980s, even moderately positive productivity growth sustained over a 25-year period would be a major accomplishment. Slower progress with policy reform in non-member economies would generate lower growth, on a par with that observed over the past 25 years (around 4 per cent per annum).

These growth assumptions have been used as an input in dynamic computable general equilibrium model to simulate implications for trade and other important variables. The features of this model and the database have been described in detail elsewhere (OECD Development Centre, 1997). The parameters used in the model as well as assumptions regarding, in particular, technological progress were derived from a review of the literature as well as on the basis of expert opinion available in the OECD Secretariat. Major assumptions include:

- a reduction of trade and transport margins by 1 per cent p.a.;
- improvement of energy efficiency by 1 per cent p.a. in OECD, 2 per cent p.a. elsewhere;
- agricultural productivity increases of 1.5 per cent p.a.
- land supply constraint; and
- oil and gas depletion modules.

4.2 Outcomes

A high growth scenario would result in substantially improved living standards across all regions and some convergence of non-members' per capita incomes towards OECD levels (figure 2).

For example, with China and India both starting from below one-tenth of the OECD average in 1995, the former's per capita income (measured in purchasing power) would rise to one-third and the latter's to one-fifth the OECD average by 2020. In general, per capita levels in non-member economies would more than double, reaching 32 per cent of the OECD average by 2020, with some countries in East and Southeast Asia exceeding the average OECD per capita income of 1995. Conversely, the world would experience much less income convergence under a low growth scenario.

Figure 2:   Income per Capita (in constant 1992 PPP)

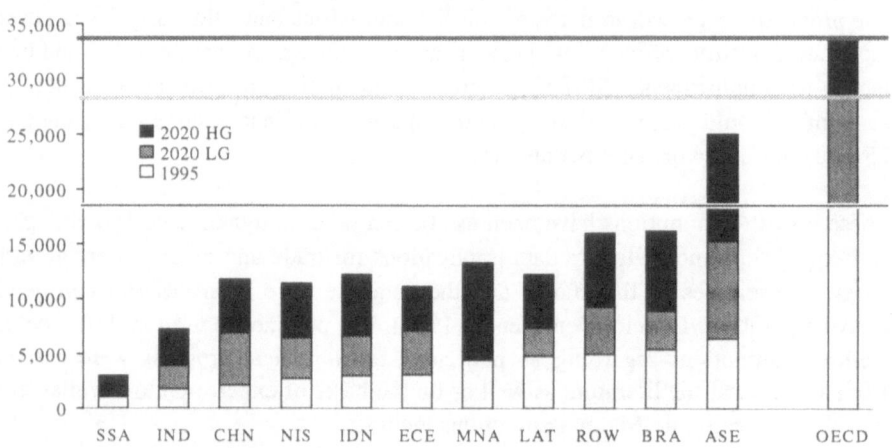

Note: Horizontal lines represent the OECD per capita income average.

Source: Based on growth assumptions and UN population projections underlying the scenarios

In a fully liberalised world economy, the shares in global output would also shift considerably in favour of non-member economies (figure 3). With world GDP increasing from 32 trillion in 1995 to 107 trillion in 2020 (in 1992 US$ and PPP exchange rates) the world economy would become tripolar, i.e. the OECD, the Big 5 and the group of other countries would hold almost equal shares in global production. In this scenario, China would become by far the largest single economy by 2020, representing slightly less than half of the OECD GDP.

With the trade liberalisation assumed in the high growth scenario, trade would rise as a share of world GDP from 30 per cent at present to about 45 per cent in 2020 (figure 4). OECD trade would expand somewhat more slowly than non-member trade, as the removal of already low tariffs in the former would provide less of a stimulus. Half of the increase in world trade would consist of trade between OECD countries and the rest of the world (figure 5). The OECD countries could expect an improvement in their terms of trade as prices of agricultural products and consumer goods, notably textiles and apparel, would fall relative to those of their major exports, notably capital goods. With the removal of tariffs on capital goods and high rates of investment in non-member economies, OECD capital goods exports to those countries would increase almost fivefold (figure 6). At the same time, consumer goods producers in non-member economies would benefit from fast-growing imports into OECD countries, with the three biggest Asian countries expanding their exports sixfold.

Figure 3:   Shares in World GDP, 1995 and 2020

(in 92 US$, using market exchange rates)

1995	2020 HG	2020 LG
25 trillion	64 trillion	47 trillion

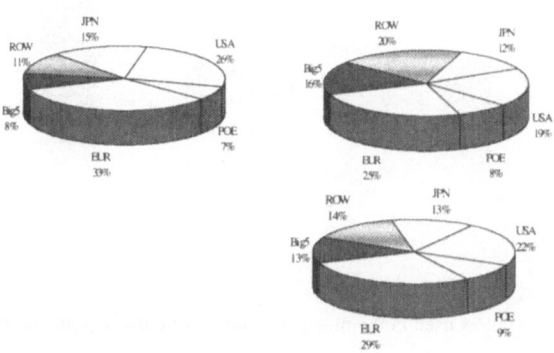

(in 92US$, using PPP exchange rates)

1995	2020 HG	2020 LG
31 trillion	106 trillion	70 trillion

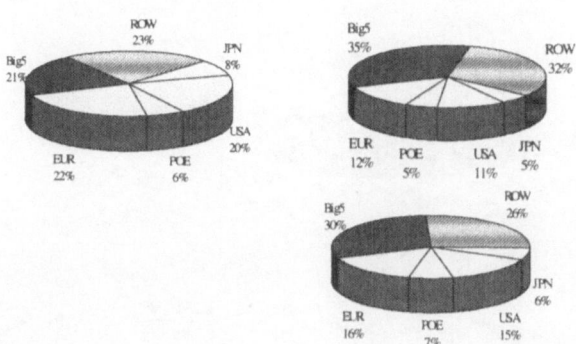

Source: Based on growth assumptions for HG and LG scenario, OECD Linkage Model

Figure 4:   Trade to GDP Ratios in 1995 and 2020 (in percent)

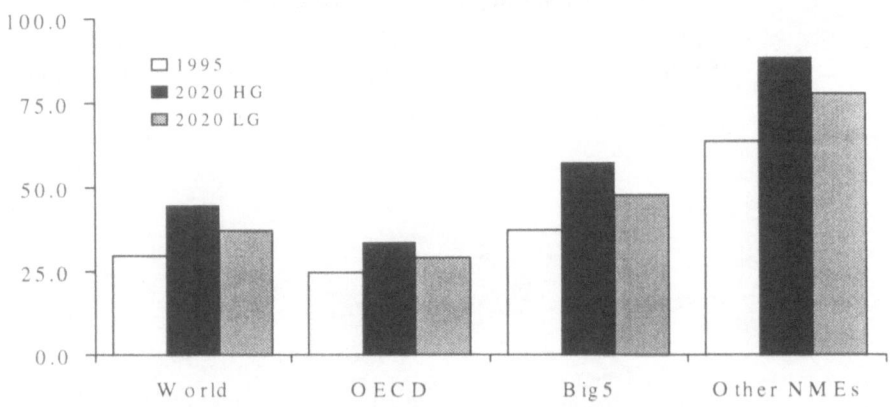

Note: For data reasons, the trade figures used in the model exclude most intra-regional trade.

Source: OECD Linkage Model

Figure 5:   Regional Patterns of Trade (in 92 US$ billions)

1995	2020 HG	2020 LG
3,600	12,600	8,300

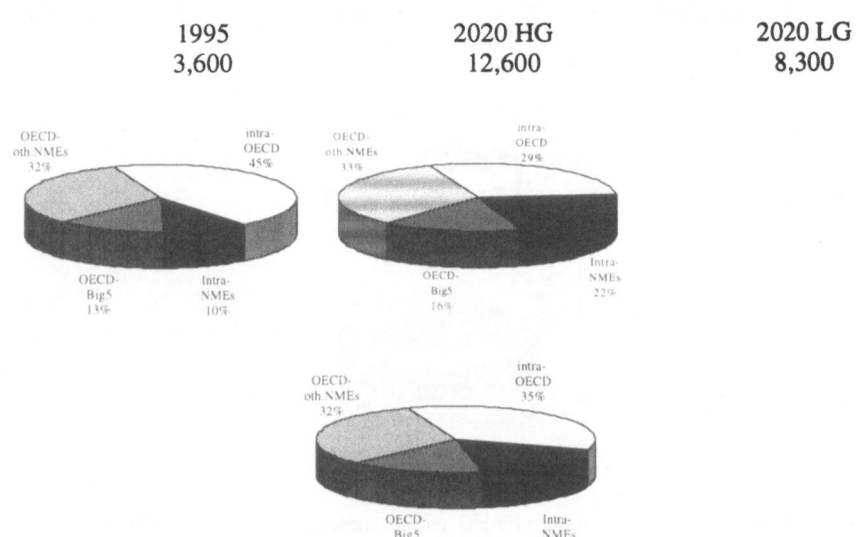

Source: OECD Linkage Model.

Figure 6:   Export Structure by Sector, 1995-2020 (1995 = 100)

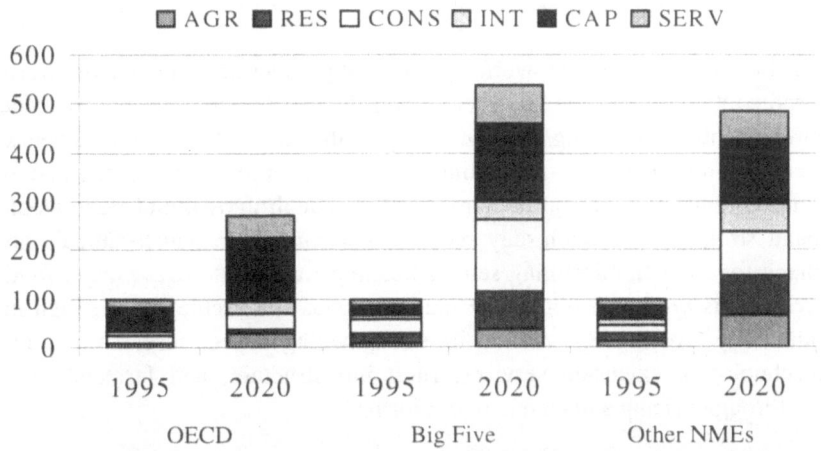

Note: AGR: Agriculture and food processing, RES: Coal, Gas, Oil and other mining, CONS:
         Consumer goods, INT: Intermediary products, CAP: Capital goods, SERV: Services.

Source: OECD Linkage Model

While non-member countries would increase their share in global output across all sectors, the relative increase would be largest in agriculture and consumer goods, where the non-OECD share of world production would rise to more than half in 2020. In the former case the growing output would largely satisfy a rising domestic demand, notably in the Big Five (the same is true of fossil fuels, notably coal in China and India), while in the latter a much larger share of production would be exported, principally to finance the non-member countries' imports of capital goods. Capital goods produced in the OECD area would still account for two-thirds of the world supply in 2020.

How would such changes in production patterns impact OECD labour markets? Could they be expected to exacerbate or ameliorate the rather adverse trends in wages and employment of low-skilled workers that have been observed over the past decade and a half? While the model simulations cannot provide a definitive answer, they indicate that: (a) globalisation could have some further impacts on wage differentials but a rather modest one, (b) this could be offset by an increase in the proportion of skilled workers in the labour force (e.g., through raising education levels) that is relatively small by comparison with recent historical experience; (c) offsetting any effects of further skill-biased technical change would pose a some-what greater challenge, requiring larger investments in education and training; (d) increasing protection would be a counterproductive response since it would slow growth of real earnings for both low-skilled and high-skilled workers.

5.   Implications

5.1   Resource Availability

In the high growth scenario, world agricultural production would expand at roughly the same rate over the next 25 years as over the past two decades, with productivity improvements contributing most of that growth. The bulk of the incremental food demand in non-members - including in large economies like China and India - would continue to be supplied domestically, though agricultural trade would also expand strongly. The doomsday scenario - according to which China's growing demand for food and declining self-sufficiency ratio would cause major food price increases - is *not* borne out by the model results. Still, achieving the high implied productivity growth rates will require strengthening of crop research (including on biotechnology), extension services, rural infrastructure, and farmers' incentives (e.g., through pricing and land tenure reforms).

Growing agricultural imports into once heavily protected markets - notably, the EU, Japan and other prosperous East Asian countries - would be supplied largely by exports from North America, Australia and Latin America (figure 7).

Figure 7:   Net Trade in Agriculture and Food Processing, HG
                 (constant 1992 US$ billions)

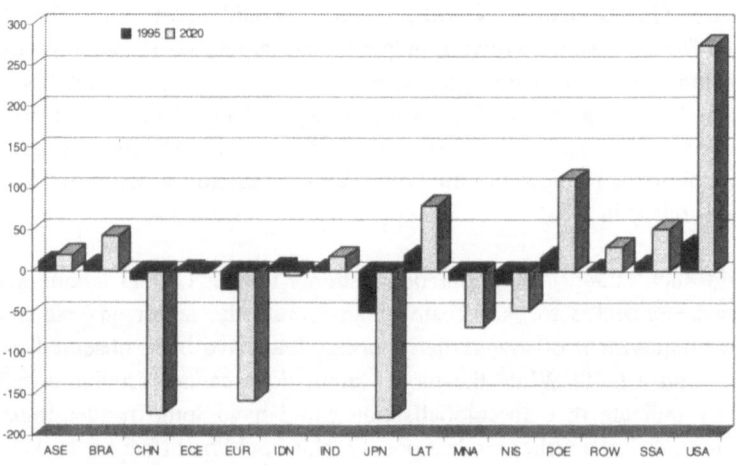

Source: OECD Linkage model.

Even with increased food trade, some regions and countries may still experience calorie deficits, but 25 years of broad-based, rapid growth in the world's poorest countries and regions would go some way to reducing the incidence of malnutrition. Energy demand would grow strongly in a high growth scenario, as the fastest output growth occurs in some of the more energy-intensive economies. Despite high projected demand growth and barring major oil or gas supply disruptions, world energy supplies should prove adequate with only moderate price increases in fossil fuels. Abundant reserves of low-cost coal in China and India would provide the primary fuel for their rapid growth, assuming transport bottlenecks can be removed. OECD oil imports are projected to rise from half to two-thirds of consumption between now and 2010. Oil importing countries will come to depend far more heavily than in the recent past on Middle East suppliers as other low-cost reserves are depleted (figure 8). Growing reliance on imports of oil (and gas) from a few major suppliers will increase vulnerability to macroeconomic shocks from supply disruptions, which could raise renewed concerns over future energy security - concerns likely to be shared by many non-member countries.

Figure 8:   Net Trade in Fossil Fuels, HG
              (constant 1992 US$ billions)

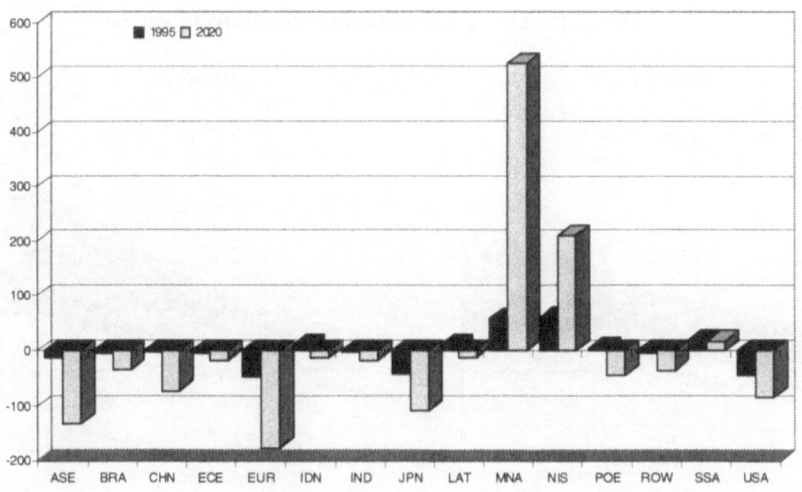

Source: OECD Linkage model

5.2 Environmental Sustainability

One important concern raised by high growth in the world economy is its environmental implications. Even with assumed energy efficiency improvements of one per cent per annum in OECD and two per cent per annum in non-OECD countries, world fossil fuel consumption could more than double in the period to 2020 and carbon dioxide (CO_2) emissions rise proportionately (figure 9). The risks of accelerated global warming and climate change may come to be perceived as unacceptably high, intensifying pressures for a more forceful global policy re-sponse. Few OECD countries are currently on track to meet the non-binding commitment made by industrialised countries at the Rio Earth Summit to stabilise CO_2 emissions at their 1990 levels by the year 2000. Even in a lower growth scenario, CO_2 emissions would still rise by about 60 per cent between now and the year 2020 (figure 9) and, without a major shift towards zero- or low-carbon energy sources, OECD countries would need to raise energy efficiency levels quite substantially - by 3 per cent per annum or more - over the next decade to achieve that target. In both scenarios, the bulk of the growth in emissions would occur in non-member countries. Thus, industrialised countries would need to do much better than merely stabilising emissions if there is to be scope for developing countries to expand their emissions while achieving the goal of stabilising atmospheric CO_2 concentrations at levels that would prevent dangerous anthropogenic alteration of climate patterns.

Figure 9: Carbon Emissions, 1990-2020 (billion tons of carbon)

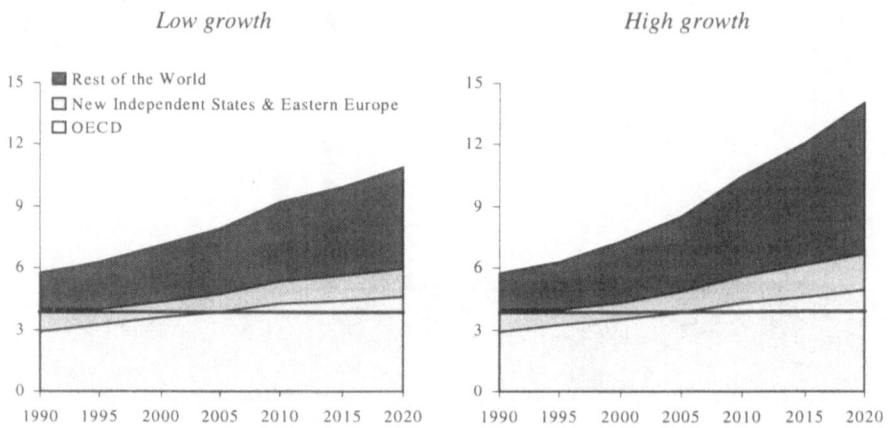

Source: OECD Linkage model.

In other environmental areas, high growth in non-member economies based on stronger linkages with the global economy could be expected to yield long-run improvements. Economic structures would shift more rapidly away from energy- and pollution-intensive sectors, cleaner technologies would diffuse more quickly, the population transition would be accelerated, and, perhaps most importantly, living standards would rise faster and with them demands for environmental quality improvements and the means to pay for them.

In response to global problems, closer international co-operation will be needed, based on acceptance of the principle of "common but differentiated responsibilities". Historically, industrialised countries have borne a greater responsibility for pollution of the global commons and consumption of natural resources. Moreover, in most cases they have a greater financial and technical capacity to pursue global environmental objectives. The major challenge is to reach agreement on the practical application of the principle.

5.3 Global Capital Flows

In the past, the domestic savings and investments of the non-member economies have been closely matched. The developing world as a whole has relied on foreign savings equivalent to only 1.5 per cent of its GDP - in other words, six per cent of developing-country investment has been financed by the rich countries. There are two main reasons to suppose that this correlation will not dramatically change, even if global capital markets were to become completely free by the year 2020. First, declining fertility rates and lower young-age dependency in non-member countries will stimulate household savings, while higher returns on invested capital will stimulate corporate savings. The high growth scenario shows that the average domestic savings rate in non-OECD countries could rise by four percentage points of GDP, a modest increase compared to that experienced in East Asia over the past quarter century. Second, financial markets are sensitive to sovereign country risk and, by implication, to a country's external debt ratio and level of current account deficits, thus constraining net international indebtedness as a percentage of GDP. These factors serve to reconfirm the expectation that the non-member economies will not absorb foreign savings at a very high proportion of their output, let alone of OECD output. Thus, fears about future global capital shortages or massive net capital outflows from the OECD area seem misplaced.

Even if *net* capital flows from OECD to non-member countries are expected to remain small, there are large gains to be derived from further international portfolio diversification. Based on assumptions of high growth and a minimum 8.5 per cent of OECD pension assets invested in the non-OECD area by 2020, the real rate of return on those assets could rise by at least 7.2 per cent. Because of the gradual build-up of assets invested in emerging economies, pension benefits would rise by a more modest 2.5 per cent by 2020. Those countries that diversify their assets into emerging economies early would likely enjoy the largest gains.

5.4 Integrating the Poorest into the Global Economy

Fast, broad-based growth in low-income countries would do much to eliminate the most abject forms of poverty and to foster income convergence both within and across countries (Chapter 5). This would help relieve the tensions and instabilities associated with pronounced poverty and inequality that are today a permanent backdrop to political discourse. Concerns have been raised, however, about the trickle-down impact of market-driven and private sector-led growth on poverty and on backward regions. Investment in the poor, primarily through education, is an essential complement to the creation of new opportunities; other facilitating measures include enhanced access by the poor to other productive assets -land and capital - through broader and more secure ownership of land, encouragement of micro-investment lending institutions. In addition, decentralised infrastructure programmes serve to link backward areas more closely to national and international markets.

The OECD countries have high stakes in the economic revival of the least developed countries and regions, since it would reduce humanitarian crises. Although OECD governments are struggling to live within tighter budgets and development assistance programmes have been among the hardest hit by budget cuts, development assistance will continue to play a valuable role in creating the conditions for integrating these countries into the global economy.

6. Summary

The model-based results, combined with other evidence, can be summarised as follows. In the event that the world economy were to realise high growth over the next 25 years, there would not likely be an agriculture or energy problem, there could well be an environment problem, and widening wage inequalities in OECD labour markets could remain a challenge.

The world should be able to produce enough food to feed the population projected for 2020, and most food needs of large economies like China and India will continue to be met from domestic production. Agricultural trade will grow, particularly between the Americas and Australasia on the one hand, and Europe and East Asia on the other. Africa will face a challenge to increase agricultural produc-tion at the rate implied by the high growth scenario. No more than today will the global adequacy of calorie supply ensure the adequacy of calorie consumption by all the world's peoples, but twenty-five years of broad-based, rapid growth in the world's poorest countries and regions would go some way to reducing the incidence of malnutrition.

Energy supplies will likely be adequate to fuel rapid economic growth over the next 25 years without drastic increases in fossil fuel prices. Dependency on Middle East oil suppliers will, however, increase significantly as other low-cost reserves are depleted. Added to this is a dependency of some OECD countries on gas imports from a few countries. This could raise renewed concerns over future energy secu-rity, concerns likely to be shared by many non-member countries.

The stresses of rapid global growth on the environment are cause for concern in the next 25 years. Business as usual could result in a doubling of global CO_2 emissions. The slow progress of OECD countries in the five years since the Rio Earth Summit towards meeting their voluntary stabilisation commitments (agreed to in the FCCC) does not provide grounds for optimism. The relatively favourable energy scenario may make for a less favourable environmental scenario insofar as future fossil fuel prices may provide only weak incentives for energy efficiency improvements and for greater reliance on renewables. Still, there remains much scope for further improvements on this score through removal of fossil fuel subsidies, in Member and non-member countries alike. Subsidy reform will need to be one part of a multipronged strategy involving a mix of instruments if the task of cutting global greenhouse gas emissions is to be effectively addressed.

With regard to earnings and employment, the model simulations and empirical evidence suggest that the impact of globalisation alone on low-skilled to high-skilled wage differentials would be small compared to that stemming from further skill-biased technical change. A moderate improvement in the skill mix of the labour force in OECD economies could counterbalance the globalisation effects as well as a sizeable portion of the effect from technical change. In a high growth scenario, low-skilled workers would enjoy an absolute improvement in their real earnings. Slower trade liberalisation as under the low growth scenario would result in a smaller rise in real earnings for both low- and high-skilled workers.

World average per capita income would be one-third less with low growth than with high growth. Poor countries, and poor regions within other countries, would make much less progress towards eliminating poverty. Energy demand would grow more slowly, as would CO_2 emissions, but the latter would still pose a major environmental challenge in the coming years. Also, with slower economic progress in non-member countries, the conditions for addressing environmental concerns, whether local, regional or global, would be that much less favourable. Economic structures would shift more slowly towards less polluting industries, cleaner technologies would diffuse less quickly, the population transition would be slowed, and, perhaps most important, living standards would rise more slowly and with them demands for environmental improvements. Likewise, with persistent poverty and income inequalities, there would be less chance of progress towards enhancing global and OECD security.

References

OECD (1997), The World in 2020: Towards a New Global Age, November, Paris.

OECD Development Centre (1997), The Linkage Model: A Technical Note, OECD Development Centre Technical Papers, Paris (forthcoming).

ANNEX 1

Regional Concordance for the LINKAGE Model

1 ASE Other East Asia
Chinese Taipei, Malaysia, Philippines, Singapore, Thailand,

2 BRA Brazil

3 ECE Eastern and Central Europe
Albania, Bulow growtharia, Czech Republic, Hungary, Poland, Romania, Slovakia, Slovenia

4 CHN China and Hong Kong

5 EUR European Union (15) plus EFTA countries
Austria, Below growthium, Denmark, Finland, France, Germany, Greece, Iceland, Ireland, Italy, Luxembourg, Netherlands, Norway, Portugal, Spain, Sweden, Switzerland, United Kingdom

6 IDN Indonesia

7 IND India

8 JPN Japan

9 LAT Rest of Latin America
Antigua & Barbuda, Bahamas, Barbados, Belize, Costa Rica, Cuba, Dominica, Dominican Republic, El Salvador, Grenada, Guatemala, Haiti, Honduras, Jamaica, Nicaragua, Panama, St. Kitts & Nevis, St. Lucia, St. Vincent, Trinidad & Tobago, Argentina, Chile, Bolivia, Colombia, Ecuador, Guyana, Paraguay, Peru, Suriname, Uruguay, Venezuela

10 MNA Middle East and Northern Africa
Alow growtheria, Bahrain, Egypt, Iran, Iraq, Israel, Jordan, Kuwait, Lebanon, Libya, Morocco, Oman, Qatar, Saudi Arabia, Syrian Arab Republic, Tunisia, United Arab Emirates, Yemen Arab Republic

11 NIS Newly Independent States
Armenia, Azerbaijan, Belarus, Estonia, Georgia, Kazakhstan, Kyrgyz Republic, Latvia, Lithuania, Moldova, Russian Federation, Tajikistan, Turkmenistan, Ukraine, Uzbekistan

12 POE Pacific OECD
Australia, Canada, Korea, Mexico, New Zealand

13 ROW Rest of the World
Bangladesh, Bhutan, Maldives, Nepal, Pakistan, Sri Lanka, Afghanistan, Albania, Andorra, Bosnia- Herzegovina, Brunei, Cambodia, Croatia, Cyprus, Fiji, Kiribati, Laos, Liechtenstein, Macedonia [former Yugoslav Republic of], Malta, Monaco, Mongolia, Myanmar, Nauru, North Korea, Papua New Guinea, San Marino, Solomon Islands, Tonga, Turkey, Tuvalu, Vanuatu, Vietnam, Western Samoa, Yugoslavia [Serbia and Montenegro]

14 SSA Sub Saharan Africa
Angola, Benin, Botswana, Burkina Faso, Burundi, Cameroon, Cape Verde, Central African Republic, Chad, Comoros, Congo, Côte d'Ivoire, Djibouti, Equatorial Guinea, Eritrea, Ethiopia, Gabon, Gambia, Ghana, Guinea, Guinea-Bissau, Kenya, Lesotho, Liberia, Madagascar, Malawi, Mali, Mauritania, Mauritius, Mozambique, Namibia, Niger, Nigeria, Rwanda, Sao Tome & Principe, Senegal, Seychelles Islands, Sierra Leone, Somalia, South Africa, Sudan, Swaziland, Tanzania, Togo, Uganda, Zaïre, Zambia, Zimbabwe

15 USA United States of America

ANNEX 2

GDP growth rate (Average annual growth in percent)

	1996-2000 HG	1996-2000 LG	2001-2010 HG	2001-2010 LG	2011-2020 HG	2011-2020 LG	1995-2020 HG	1995-2020 LG
ASE	7.7	6.1	7.0	4.8	6.4	4.2	6.9	4.8
BRA	5.4	3.8	6.1	3.0	5.1	2.8	5.6	3.1
CHN	9.3	7.9	8.2	5.3	7.2	4.8	8.0	5.6
ECE	5.5	2.0	5.5	3.8	4.0	2.7	4.9	3.0
EUR	2.4	2.4	2.7	2.0	2.1	1.3	2.4	1.8
IDN	7.5	5.9	7.0	4.1	6.7	4.0	7.0	4.4
IND	6.5	4.4	7.2	4.3	6.6	4.2	6.8	4.3
JPN	3.3	3.3	2.9	2.0	2.3	1.2	2.7	1.9
LAT	4.3	3.0	5.9	3.2	5.1	3.1	5.3	3.1
MNA	5.0	2.1	7.1	2.2	6.9	2.2	6.6	2.2
NIS	3.5	1.1	6.0	4.2	6.9	4.0	5.8	3.5
POE	4.3	4.3	4.7	4.0	4.3	3.4	4.5	3.8
ROW	6.5	5.0	6.6	4.3	6.5	4.0	6.5	4.3
SSA	4.6	2.8	5.0	2.8	5.8	2.6	5.2	2.7
USA	2.2	2.2	2.7	2.1	2.6	1.5	2.6	1.9
Total	3.3	2.9	3.9	2.6	3.7	2.1	3.5	2.4
OECD	*2.7*	*2.7*	*3.0*	*2.2*	*2.6*	*1.6*	*2.8*	*2.1*
Non-OECD	*6.0*	*4.1*	*6.7*	*3.9*	*6.3*	*3.7*	*6.4*	*3.9*
Big5	*6.5*	*4.7*	*7.1*	*4.4*	*6.6*	*4.1*	*6.8*	*4.3*
Other NMEs	*5.7*	*3.7*	*6.4*	*3.6*	*6.1*	*3.3*	*6.1*	*3.5*

Note: HG: High Growth, LG: Low Growth

Source: OECD Economics Department and Development Centre

Population growth rate (Average annual growth in percent)

	1996-2000	2001-2010	2011-2020	1995-2020
ASE	1.54	1.29	0.95	1.20
BRA	1.25	1.15	0.95	1.09
CHN	0.90	0.67	0.60	0.69
ECE	-0.12	-0.08	-0.11	-0.10
EUR	0.21	0.02	-0.09	0.02
IDN	1.49	1.20	0.98	1.17
IND	1.62	1.36	0.99	1.26
JPN	0.22	0.05	-0.26	-0.04
LAT	1.73	1.53	1.27	1.47
MNA	2.40	2.28	1.85	2.13
NIS	0.01	0.07	0.05	0.05
POE	1.27	1.05	0.82	1.00
ROW	2.09	1.80	1.49	1.74
SSA	2.79	2.68	2.45	2.61
USA	0.79	0.74	0.76	0.75
Total	**1.38**	**1.24**	**1.08**	**1.21**
OECD	*0.58*	*0.44*	*0.34*	*0.43*
Non-OECD	*1.54*	*1.39*	*1.21*	*1.35*
Big5	*1.11*	*0.92*	*0.74*	*0.89*
Oth. NMEs	*2.16*	*2.01*	*1.76*	*1.94*

Source: Population Projections, United Nations, 1996.

Globalisation Implications for Industrial Relations

Sylvia Ostry

1.   Introduction

Linkages among countries or the increasing integration of the global economy has proceeded in phases since the end of World War II. The "shallow integration" of the early post-war decades was focused mainly on border barriers to trade flows. Reducing these barriers through successive rounds of GATT negotiations clearly involved some degree of constraint on governmental freedom of action but it was limited. Indeed the original concept underlying the GATT was to maintain a balance between domestic priorities and international obligations.

Global integration began to intensify in the 1970's and 1980's when the liberalisation of capital flows and deregulation of financial markets steadily eroded the scope for national monetary and fiscal policies. But by the end of the 1980's a new phase of interdependence had begun, led by a marked increase in foreign direct investment primarily in capital- and technology-intensive sectors and services. With international rivalry intensifying, global integration of production will grow as firms use the new information and communication technologies (ICT) to capture the economies of scale and scope stemming from geographic diversification. This deepening integration of the economy has shifted the focus of trade policy inside the border - as exemplified by the negotiations on services or intellectual property in the Uruguay Round.

The third phase of integration, to which the term of globalisation has often been affixed, was and is closely linked to the ICT revolution (and technological changes in transportation, especially containerisation) which made it cheaper and easier to manage far-flung and widely dispersed production networks. Even within service industries, rapid developments in these technologies have increased tradability and enabled firms to allocate portions of the production process to foreign affiliates. The agent of globalisation, the MNE, is the main funnel for the three engines of growth: trade; capital; technology.

The full impact of the ICT revolution clearly has some distance to go and a fourth phase of global integration is now visible in the growth of electronic commerce. Reflective of the underlying shift in the technology trajectory from "hard" to "soft", and the growing importance of "network markets", especially telecommunications, this new world of cyberspace literally eliminates borders so that the term "domestic policy" could become an oxymoron! Be that as it may, the main issue for the subject at hand is that the steadily deepening integration of the global economy is in effect creating a momentum to a global single market. Of course we may never arrive at that destination but the pressure for system convergence - harmonisation of domestic policies and institutions - will be fed by locational competition for investment; regulatory arbitrage by MNE's; rapidly changing communication modes; and international economic policy in both multilateral and regional fora.

Among the many domestic policies and institutions that are being and will continue to be subject to the pressures of globalisation, none is more fundamental (or more politically sensitive) than the industrial relations regime. The compelling arguments for a global single market are economic. Consumers would be able to buy the best products at the lowest prices anywhere and everywhere. The gains stem not only from the static, once-and-for-all efficiency gains from eliminating barriers but also, and more importantly, from dynamic efficiencies which would increase growth and create new jobs as global competition forced firms to restructure, network, and innovate. "Global" growth would increase, "global" consumers would benefit, but the distributional effects of these gains, the distribution of winners and losers both among and within countries, would affect the immobile factors of production, the land and all but the most highly skilled and educated labour of the nation state. It is, indeed the alleged effects of globalisation on labour markets which has aroused much of the recent attack on globalisation. Whatever the impact of globalisation on labour markets, that impact is mediated by the institutional arrangements of the industrial relations regimes which vary widely among the countries of the OECD.

2. Industrial Relations Regimes: Convergence or Divergence?

The demand for less-skilled and less-educated workers has weakened while that for the highly skilled has strengthened in all OECD countries since the mid-80's. This has generated a debate, still ongoing, as to the cause of this trend and until recently the favoured mainstream explanation has been technological change biased in favour of the highly skilled and highly educated, with international factors playing a minor role. Now studies have suggested that when foreign investment is taken into account the threat of delocation acts as a damper on wages. So far this effect (manifested by rising elasticities of demand for labour) has been most apparent in manufacturing but will spread to services as intra-firm global networks are established. Only the most educated and most skilled, who are also the most mobile and can therefore command the highest rewards, are likely to be immune to this globalisation effect. While new studies on these labour market developments will continue to be published it could be argued that the debate about globalisation versus technology is somewhat irrelevant since, as stressed earlier, the two forces are increasingly interrelated. Trade, investment and technology are so intertwined that trying to disentangle their separate impact would be formidable. Moreover, the migration of the highly skilled - the flow of brainpower or wetware - should also be added to the equation. Unfortunately the political debates will not be concerned with the merits of alternative models. It's much easier to round up the usual suspects - foreigners, whether immigrants or traders or countries with low wages.

Another feature of the globalisation versus technology literature has been the attribution of the notably differential impact of these combined forces among OECD countries. The "stylised facts" - rising unemployment in Europe, rising inequality in the U.S. - have been largely attributed to labour market characteristics, viz. European rigidity versus American flexibility. This view of a growing transatlantic polarity (with Japan's system rightly viewed as sui generis) has become central to much of the policy debate on labour market reform in many European countries. In effect, the answer to the question "Convergence or divergence?" is convergence (to the American model) or increasing divergence in growth and jobs. Ralf Dahrendorf has summed it all up very neatly: would you rather be poor in the U.S. or unemployed in Europe?

As is usually the case in contentious policy debates, the popular "stylised facts" approach skims lightly over a range of more complex issues which emerge from a deeper probe of transatlantic differences and the impact of these differences. Thus it is true that U.S. labour markets are uniquely flexible in part (but not only) because of a unique industrial relations regime which includes both private and public institutions and cultural norms. Indeed the uniqueness of the American system is best highlighted when compared with the Canadian which is its closest counterpart (see below). However, there is also significant variation in IR regimes within Europe. Since these institutions intermediate between the domestic impact of globalisation, before presenting the evidence on convergence it's important to define what we mean by an IR regime.

An industrial relations regime comprises the institutional arrangements which govern (or significantly influence) labour market outcomes. While a range of typologies have been proposed by IR experts all include three characteristics: union density or the percentage of union members among the employed; collective bargaining patterns such as multi-employer or individual firms; and the scope of legal mechanisms which extend the terms of collective agreements to non-union workers.[1] Both union density and coverage are easily quantifiable and figure 1 demonstrates the considerable diversity of IR regimes within the OECD. It also underlines the significance of government regulation as well as bargaining patterns in extending coverage.

---

[1] Cf. Jacoby (1995); Traxler/ Turner (1991); Blanchflower/ Freeman (1992); International Labour Organisation (1997) and OECD (1997).

106

Figure 1:   Trade Union Density and Collective Bargaining
                  Coverage Rates, 1994

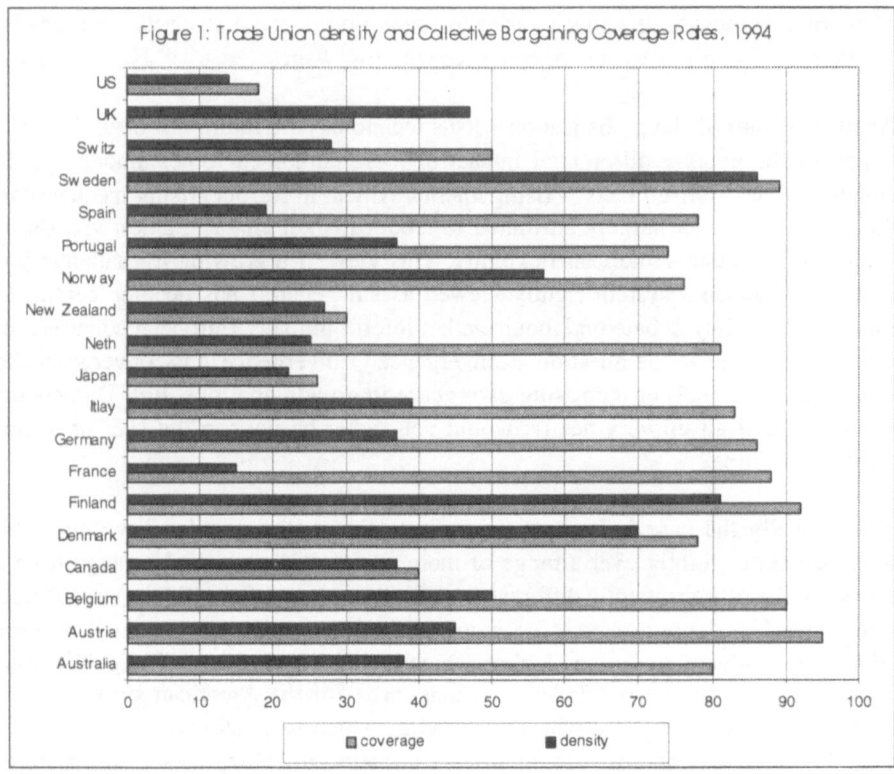

The significant differences in coverage shown in figure 1 illustrate a clustering of
countries. Coverage rates are significantly higher in continental Europe than in
North America or Japan, with the U.K. closer to the latter group. By and large these
differences, as noted above, reflect not only legally mandated extension, but also,
and more importantly, the nature of the collective bargaining arrangements, the degree of centralisation and the modes of co-ordination between the bargaining partners. Centralisation and co-ordination are linked (an exception is Japan, where enterprise bargaining is the norm, but bargaining strategies for the annual negotiations
are co-ordinated by employers associations). For the most part, however, more centralised and co-ordinated systems have far higher coverage rates. The most decentralised and uncoordinated regimes are those of the U.S. and Canada, yet there are
significant differences in union density and coverage, not only because of government legislation which is more protective of unions in Canada but also due to deep-
seated cultural and value differences.[2]

---

2 Cf. Lipset/ Meltz (1998).

Have globalisation forces generated a trend to convergence among these IR regimes? There have been a number of country studies on selected aspects of the institutions but no comprehensive surveys. As summarised by the OECD, while centralised bargaining has considerably weakened in some countries, e.g. Sweden, New Zealand, Australia and the U.K., other countries including Norway, Italy and Portugal have moved toward greater centralisation. There has, in other words, been no uniform trend in bargaining patterns. And while union density has declined in virtually all OECD countries with Spain and Sweden being the only significant exceptions this has not significantly affected coverage rates which are primarily determined by bargaining patterns and extension mechanisms. The decline in union density, it should be noted, largely reflects the shift away from manufacturing to services although in the U.S. legislation restricting union power has also played a significant role. In sum, however, there does not appear to be convergence of IR systems - at least not yet. This view accords with the findings of other recent studies.[3]

However these macro measures do not capture developments at the enterprise level where changes in enterprise organisation and industrial relations arrangements have been spreading rapidly. This is often described as a shift from Taylorism to Toyotism (or lean production) and the new industrial relations structures, which involve cross-functional teams and multi-skilling, are known generically as Human Resource Management (HRM). In Japan and many European countries the new HRM structures have been introduced with trade union collaboration. But in the U.S. and the U.K. the process has been more confrontational.[4] The most recent ILO annual report says that these developments can be interpreted in two ways: "either the union facilitates the introduction of these methods, or they are used to counter the influence of unions" ... and where unions play a consultative role in decision-making "the new practises expand more rapidly."[5] Thus, at the same time as the broad institutional architecture may be slow to change, more scope for institutional innovation at the enterprise level can provide a means of adjusting to globalisation. If this is the case then a more fundamental question arises: which IR regimes are more adaptable? Unfortunately at present this very important question can't be answered. In the absence of more comprehensive micro data one can only note that firm-level case studies underline the view that there is "no one size fits all" model of institutional governance but considerable variation by industry sector and over time.[6] However, more co-operative approaches between management and union to

---

[3] Cf. Ehrenberg (1994); ILO (1997).

[4] ILO (1997).

[5] ILO (1997); Locke/ Kochan/ Piore (1995).

[6] Cf. Traxler/ Unger (1994).

implement organisational change and new work arrangements do seem to work better than adversarial models in facilitating technological change, worker training, and enterprise innovation.[7]

There are thus, as yet, few signs of convergence of IR regimes among OECD countries - workers continue to work under different rules, although those rules may be in process of changing at the enterprise level. So what does this mean for labour market outcomes in these countries? The July 1997 OECD Employment Outlook provides the most comprehensive summary of current research on the issue and the conclusions are worth quoting:

"While it is premature to draw definitive conclusions on this issue, the evidence presented in this chapter does not show many statistically significant relationships between most measures of economic performance and collective bargaining. This negative conclusion holds irrespective of whether collective bargaining systems are proxied by measures of trade union density, collective bargaining coverage or the centralisation and co-ordination of bargaining. One exception to these negative findings is that there is a fairly robust relation between cross-country differences in earnings inequality and bargaining structures. More centralised/co-ordinated economies have significantly less earnings inequality compared with more decentralised/uncoordinated ones. In addition, while not always statistically significant, the chapter finds some tendency for more centralised/co-ordinated bargaining systems to have lower unemployment and higher employment rates compared with other, less centralised/co-ordinated systems."[8]

Figure 2:   Alternative Incidence Measures for Low-Paid Employment, 1986 - 1991

---

[7] Cf. Traxler/ Unger (1994).

[8] OECD (1997).

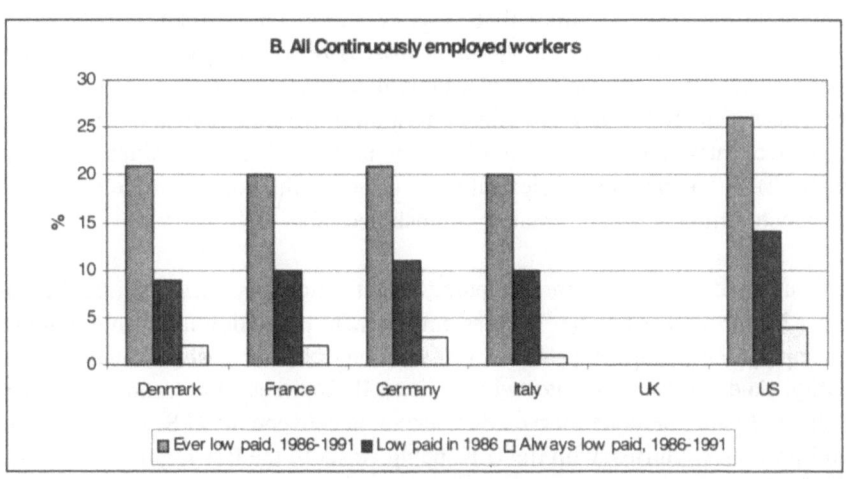

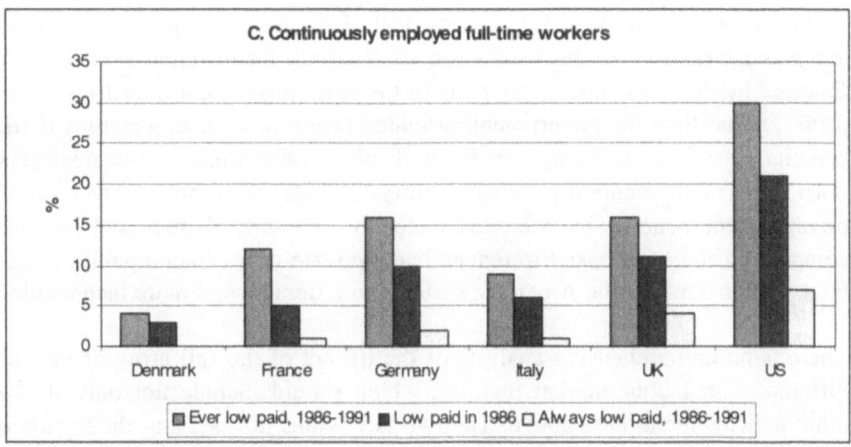

Source: OECD Employment Outlook

Note: (A) Low pay is defined as bottom quintile of weekly/monthly earnings of all full-time wor-
    kers; (B) Low pay defined as bottom quintal of annual earnings of all workers; (c) low
    pay defined as below 0.65 median weekly/ monthly earnings of all full time workers

The association between IR regimes and inequality accords with the findings of a
more recent study of the United States, Canada and France.[9] Over the 1980's, relative wages of unskilled workers fell the most in the U.S., somewhat less in Canada,
and not at all in France - consistent with the flexible-rigid transatlantic paradigm.

_____

[9] Cf. Card/ Kramarz/ Lemieux (1996).

But, as in the OECD studies, there was no significant correlation between downward wage flexibility and employment loss. However employment loss should be distinguished from unemployment which, of course, also reflects job creation. The American economy has a far higher proportion of low-paid jobs than any other OECD country and this is not offset by higher lifetime earnings mobility (see figure 2).[10] The downward flexibility of wages at the bottom in the U.S. may well be a factor influencing the creation of unskilled jobs.

Labour market outcomes are, of course, affected not only by IR regimes but also by other labour market policies such as employment protection laws; minimum wages; unemployment insurance; the level of non-wage labour costs to finance social security. And the two are intertwined, since IR regimes affect and are affected by politics. Again, there are marked differences as between the U.S. and most continental European countries, with the U.K. being closer to the less regulated U.S. There is a particularly striking transatlantic difference in the size of non-wage labour costs to finance social security (e.g. well over 40% in most continental European countries versus under 30% in the U.S.). A good deal of the American post-war welfare system was internalised in the 1950's and 1960's (and, for different reasons in Japan) financed by the quasi-monopoly rents in the mass production industries. The companies, rather than the government, provided health insurance, generous retirement benefits, vacations, training, etc. for both blue collar workers and managers. In other parts of the economy social security benefits are minimal. Over time these private sector benefits have eroded under the pressure of increased international competition since it's easier to renegotiate a private rather than a public social contract. And this makes the American system more flexible and more inequitable.

There is no comprehensive analysis of the impact of the full array of institutional differences in labour market regimes, which should include not only IR but the other policies mentioned above. The two key issues in assessing the impact on labour costs are the extent to which costs are shifted from employers to workers and the degree of enforcement of legislated rules. Another significant policy "escape hatch" is the exchange rate, which was a key mechanism of adjustment in the deepening integration of the North American economies, especially Canada and the U.S. Further, not only costs but benefits differ among different regimes. These would best be captured by firm level productivity performance which does not appear to be correlated with the presence or absence of unions.[11] More broadly, the limited studies which are available do not suggest significant correlation between labour market regulations and labour costs.[12]

---

[10] OECD (1997).

[11] Cf. Traxler/ Unger (1994).

[12] Cf. OECD (1996); Ehrenberg (1994); Blank in Freeman (1994).

Nonetheless, there is a widespread view that the European systems (excluding the U.K.) support the jobs of prime age males - the "insiders" - at the expense of the "outsiders", primarily women, youth and older men. From this vantage point the persistently high levels of European unemployment, as compared with Japan and the U.S., result from the greater reluctance of European firms to hire new workers, preferring to adjust to external pressure by varying hours of work rather than increasing employment.[13] The evidence supporting the insider-outsider hypothesis thus far is tentative. More importantly, the slower rate of job creation in Europe doesn't necessarily arise solely or even primarily from labour market rigidities. In the remainder of this paper I want to look at system differences in a broader sense.

3.   Markets and Non Market Systems

A number of recent studies have suggested that product-market barriers are significant impediments to job creation in Europe. These regulatory barriers affect mainly the service sectors, including retailing, financial services and telecommunications.[14] Another significant barrier to job creation stems from constraints on the formation of new firms, whether these take the form of regulatory impediments or lack of capital. These constraints would reduce the demand for labour by reducing the supply of entrepreneurs.[15] These studies are illustrative of a more fundamental issue, i.e. that a focus solely on IR regimes as the primary determinant of labour market outcomes is unlikely to produce meaningful policy guidance. These regimes are embedded in an institutional system which includes not only labour markets, but product and financial markets, as well as non-market institutions such as unions or components of innovation systems such as universities, etc., etc. The challenge for policy analysis, it need hardly be said, is formidable. Institutions do not exist in neo-classical models and recent developments in incorporating some institutional elements into mainstream economics, while promising, are still very preliminary.

This being said, stylised facts can be helpful in highlighting systems. The transatlantic differences in labour market maladies reflect fundamental systemic differences between Europe and the U.S. The American paradigm - fluid, flexible and disposable - with a far greater tolerance for inequality and far greater scepticism (or hostility) to government, is deeply rooted in America's historical origins. The continental European model (and I agree generalisation may be risky) includes a far more extensive welfare state inherited from the post-war period and indeed created largely because of the bitter experience of the inter-war years. The contrast between

---

[13] Cf. Flanagan in Büchtemann (1993); Andersen (1997).

[14] Cf. McKinsey Global Institute case studies in Krueger and Pischke (1997).

[15] Cf. for theoretical model, see Ibid.

the U.S. and Europe has been characterised as a contrast between Exit and Voice.[16] An Exit paradigm is far more adaptable in a period of rapid change because in Exit social change is ensured through an anonymous mechanism that ensures victory for the most efficient - winners are rewarded and losers appear to disappear. But when losers have a Voice and governments (and other institutions) must engage in a long and difficult process of political renegotiation of the post-war social contract, the "rigidities" which were not terribly important when growth was high, and moving in established channels, become powerful impediments to adaptability in a time of ongoing and pervasive transformation.

The unique character of the American system is particularly consonant with a key aspect of the current environment, i.e. the ongoing revolution in information and communication technologies which is propelling a massive restructuring of the economy. As in similar periods in the past, the result will be a long-term increase in living standards but lots of losers en route to that new golden age. The major source of jobs today and for the foreseeable future will be in the services sector while the shake-out continues in manufacturing industries. The American system is especially hospitable to the generation of new products and new markets in software, communications, and financial services. Indeed the shift of the ICT technology trajectory from hard to soft amplifies the advantage of the English language and indeed, the advantage of Hollywood, since a major driver of the technology is entertainment.[17] Thus although computer production and commodity chips have moved offshore, lost jobs have been replaced by the growth of the software and content sectors and by retail stores that sell computers. Most jobs in the American computer industry are now in R&D, design, engineering, customer support, i.e. highly skilled, well paid, upper-end jobs. Even in software, where computer-aided software engineering (CASE) has replaced many engineers and programmers and is moving offshore, the demand for "the creative, unstructured, problem-solving tasks that comprise the artistic heart of software engineering"[18] has increased. Since this talent is in short supply, a growing number of the skilled professionals in the U.S. are immigrants - a brain drain that is likely to increase. Clearly the rapid pace of change in the new technological revolution requires a continuing, or even accelerating, pace of structural adaptation.

There are a number of other characteristics of the ICT trajectory that are strikingly concordant with the American system. Unlike previous technological revolutions, information technology which rests on the codification and storage of information is affecting services much more profoundly than other sectors. Services such as financial; telecommunications; consulting and auditing, are more regulated than

---

[16] Cf. Hirschman (1993).

[17] Cf. Judy/ D'Amico (1997).

[18] Ibid., p. 20.

manufacturing and thus regulatory reform is an essential element in promoting innovation. The American telecommunications sector, for example, achieved "first mover" advantage by moving first on regulatory reform. Global economies of scale in these markets vastly amplify the value of very small differences in talent or ability - the "winner takes all" syndrome.[19] Thus first mover advantage in telecommunications and multimedia is reinforced by positive feedback links to magnify the income of superstars, whether entertainers or lawyers or brilliant scientists.

Further, it's easier to codify information to be embodied in material goods, routine management tasks, or commoditized software than the soft skills that require talent and creativity or the low level jobs in retail, health care, domestic service, gardening, janitorial work, etc. which are likely to increase significantly because of demographic change. More of these low level jobs are likely created if wages can be kept low or pushed down. Thus there is an inherent tendency to widening inequality in this technological paradigm stemming from growth at both the high and low end of the income scale. While it would be wrong to say that all Americans ignore the growing disparities in income - or do not fear the costs in terms of declining social cohesion - it is certainly agruable that the popular tolerance for inequality is greater than in other OECD countries. This, once again, comes through clearly in comparative analysis of Canada and the U.S., whose economies and societies have been so intertwined over the past century or more. Thus even if a tolerance for inequality is a strategic asset in the new innovation paradigm - and assuming the benefits outweigh the costs - it is not an asset easily replicated in different systems. Convergence seems unlikely, even if it were desirable.

4.   Conclusions

The main theme of this paper was to explore the impact of globalisation on IR regimes in the OECD. I have concentrated on the European–North American comparison since this has been a key focus of policy debate both in Europe and the U.S. with Canada, as always, somewhere in between.

For those seeking clear policy recipes the paper will remind them of the old joke about economists: when an economist says there's no quick fix it's because he (or she) suspects there's no long-term solution. But that was not the paper's intent. The review of IR regimes and labour market policies does not provide empirical support to the flexibility vs. rigidity model. Nor is there meaningful evidence of convergence in IR regimes. To the contrary, the transatlantic divergence is widening. But so is the divergence in job creation. And that really is the heart of the matter.

---

19 Cf. Rosen (1981); Rank/ Cook (1995).

The paper argues that IR regimes are embedded in far broader and deeper institutional systems. System convergence is neither feasible nor desirable. But the paper also argues that the policy focus - on job creation - has to go well beyond labour markets. The design of the policy template should be on identifying and eliminating impediments to job creation in all markets and to enhance incentives to innovation, the source of new jobs and new markets. Two other issues deserve more consideration and more comparative research. One concerns the extent to which wage floors at the low end of the income scale inhibit the creation of jobs in the services sector, especially personal services and retailing. If this proves to be a significant barrier, then the question of the trade-off between jobs and equity will have to be considered. The other issue concerns the increased mobility of the increasingly scarce, highest quality, human capital - the brain drain. Since the basic building block of all innovation systems is human capital I would hope that better data and more analysis of these flows will be a central theme in innovation studies.

References

Andersen, T., et.al.: "Make Room for Outsiders," *Financial Times*, November 20, 1997.

Blanchflower, D.G., Freeman R.B.: "Unionism in the United States and other Advanced OECD Countries," *Industrial and Labor Relations Review*, Vol. 31, No. 1, Winter 1992, pp. 56-77.

Card, D., Kramarz, F., Lemieux,T.: *Changes in the relative structure of wages and employment: comparison of the U.S., Canada and France*, NBER Working Paper 5487, March 1996.

Ehrenberg, R.G.: *Labor Markets and Integrating National Economies*, Brookings Institution, Washington D.C., 1994. For trade union density data, 1985-1995, see ILO, op.cit.

Flanagan, R.J.: "Hiring behaviour and unemployment," in C. Büchtemann (ed.), *Employment Security and Labour Markets Behaviour: Interdisciplinary Approaches and International Evidence*, ILR Press, Ithaca, N.Y., 1993.

Freeman, R.B., (ed.): *Working Under Different Rules*, A National Bureau of Economic Research Report, New York, 1994.

Hirschman, A.: *Exit, Voice and Loyalty*, Cambridge, 1971.

International Labour Organisation (ILO), *World Labour Report: Industrial Relations, Democracy and Social Stability*, Chapter 5, Geneva, November 1997.

Jacoby, S.M., (ed.): *The Workers of Nations: Industrial Relations in a Global Economy*, Oxford University Press, New York, 1995.

Jacquemin, A., Wright, D.: "Corporate Strategies and European Challenges Post-1992," Journal of Common Market Studies, December 1993.

Judy, R.W., D'Amico, C.: *Workforce 2020: Work and Workers in the 21st Century*, Hudson Institute, Indianapolis, 1997.

Krueger, A.B., Pischke, J.S.: Observations and Conjectures on the U.S. Employment Miracle, NBER Working Paper No. 6146, Aug. 1997.

Lipset, S.M., Meltz, N.M.: "Canadian and American Attitudes Toward Work and Institutions," *Perspectives on Work*, Vol. 1, No. 3, 1998.

Locke, R., Kochan, T., Piore, M.: *Employment Relations in a Changing World*, MIT Press, Cambridge, Mass., 1995.

OECD *Employment Outlook*, July 1996.

OECD *Employment Outlook*, Paris, July 1997.

Rank, R.H., Cook, P.J.: *The Winner-Take-All Society*, The Free Press, New York, 1995.

Rosen, S.: "The Economics of Superstars," *American Economic Review*, Volume 71, No. 5, December 1981.

Traxler, F., Unger, B.: "Governance, Economic Restructuring, and International Competitiveness," *Journal of Economic Issues*, Vol. XXVIII, No. 1, March 1994, pp. 1-19.

Traxler, F.: "Collective Bargaining and Industrial Change: A Case of Disorganisation?", *European Sociological Review* (forthcoming).

Turner, L.: *Democracy at Work, Changing World Markets and the Future of Labor Unions*, Cornell University Press, 1991.

European Integration as an Answer to Technoglobalism

Luc Soete

1.   Introduction

Parallel to the process of economic integration, as it has taken place over the last twenty years and particularly within the framework of the creation of the large European "Single Market", European economies have been confronted by a significant increase in the degree of structural change at the world level. The last ten years can indeed be described as a period of historic, major structural change at the world level: the collapse of the former socialist countries and their rapid opening up to market-led economic incentives; the shift in world market growth from the North Atlantic OECD area to the Pacific basin area with an increasing number of Asian economies outperforming the developed countries' growth performance; the creation of new regional trading blocks in North and South America, in Asia, in the Middle East and in Southern Africa, with more rapid growth in trade within than between such integrating trade areas; the surge in foreign direct investment in these trade blocks with large global firms aiming at presence in each of these markets; and last but not least the dramatic reduction in the costs of information and communication processing, opening up an increasing number of sectors to international trade and giving at least the impression of a dramatic reduction in physical distances -- the world as a village.

This fast-paced global restructuring process raises some fundamental policy challenges at both the national and European levels. At the national level, it has made policy makers much more aware of the increased international implications of their policy actions. Policies that might appear "sustainable" within a national or even European context, might increasingly appear less so in an international context. While the impact of opening up to global international restructuring might still be in its initial stages, it has rapidly brought to the forefront how degrees of freedom of policy actions have been dramatically reduced in a wide variety of different fields. This does not only hold for traditional macro-economic policy, but also for social policy, tax policy, social security policy and other policies traditionally preserved at the national level.

At the same time, globalisation is also raising fundamental questions with respect to Europe's own integration process. The latter is characterised by economic aims which appear increasingly overtaken in their purpose and speed of implementation by the broader world wide integration process (one may think of the recent WTO Singapore agreement on the liberalisation of information technology trade). It brings to the forefront the question of whether the old process of economic integration whereby the central aim is the reaping of the scale advantages of the large European market with its

350 million consumers, is not, at least in the area of manufactured goods, entering into its decreasing marginal return phase and is now not in need of new policy reflection and possible policy action.

In this paper I briefly discuss some of the main interactions between countries' growth dynamics and firms' innovative and technological capabilities, as they have evolved in this increasingly global environment and the challenges they raise with respect to national technology support policies. The analysis points to the complexity and historically grown importance of the science and technology institutional framework, what has been described as the national system of innovation (Freeman, 1987, Lundvall, 1992, Nelson, 1992). The effectiveness of such national institutional framework is clearly being challenged by the increasingly global behaviour of private firms; it is, however, also become challenged by the growing role and importance of regional conditions, including regional policy making, for creating and maintaining locational advantage.
The voluminous literature points to the variety of conditions both with respect to the nature of the new technology (radical, incremental) and the different sectors (science-based, process-oriented, etc.) likely to exist and their normative, national policy implications. This variety of conditions is not just reflected in an impressive variety of national institutional set-ups governing the creation and diffusion of innovation and technological capabilities, it also highlights the difficulties in identifying "best practice" policy. Hence, despite the obvious advantages to international co-operation and the identification of some "global level playing field" in this area, the scope and nature of policy competition versus harmonisation remains an issue open to debate.

In the final section I then turn to some of the new European economic integration policy challenges. These are admittedly somewhat short in practical content. At this stage the aim is really only to wet the policy maker's appetite.

2.   "Global" Firms, Technological Capabilities and Countries' Growth Dynamics

The structural change pressures described above have led to a shift in the form of globalisation. Apart from the globalisation induced by the rapid international financial liberalisation, not discussed here, globalisation appears indeed no longer simply a question of "globalising" sales with its accompanying services such as marketing, distribution and after-sales service, but involves to a greater extent production, including production of component suppliers; investment, including intangible investment; merger acquisitions, partnerships, so-called "strategic" alliances; etc.

As discussed in many recent contributions in the international business literature,[1] the aims of firms is increasingly directed towards global presence strategies which find a balance between reaping some of the scale advantages of global markets increasingly associated with intangibles (research, communication, marketing, logistics, management), yet exploiting also the often geographically determined diversity of consumers and production factors. The large multinational firm's organisational, as well as production technology will give it the necessary flexibility to confront this diversity. The decentralisation of its production units and even new product development, together with a diversification of sub-contractors, will enable it to take full advantage of this diversity. This explains a.o. the apparent contradictory "glocalisation" trend based on physical presence under what appears sometimes rather "autarchic" production conditions in the various large trade blocks (EU, NAFTA, ASEAN, China) with often highly differentiated "local" products, yet increasing global exchange of some of the core technological competencies of the firm, including the establishment of alliances, networking with other firms and other forms of international exchange of relevant information.

As made explicit in the term "multi-domestic" firm, the actual location of the firm's plant will depend heavily on local environment factors. Whereas the locational choice will often depend on the availability of local skills, infrastructure and access to knowledge, the firm itself will of course contribute to the long-term growth and availability of human resources, access to knowledge, local suppliers' know-how and networks. These often scarce and sometimes geographically "fixed" factors contribute to create the increasing return growth features of long term development.

These apparently opposing trends raise a number of important policy issues, not in the least with respect to the level at which policy should be implemented so as to be most effective. It is obvious that global or multi-domestic firms question increasingly the meaning of much national policies. In many cases such firms might behave in an as good "citizenship" manner than national firms, in other cases apparently not. It is difficult if not impossible for governments to draw lines here: the current OECD guidelines with respect to foreign direct investment provide little more than a voluntary "standard" of good international behaviour.

In the late 1980's, proposals were made to use, for national policy purposes, some notion of the "degree of globalisation" of firms, measured e.g. in terms of the composition of boards, the international breakdown of key management posts, of research laboratories and more generally physical and intangible investment. Given the ongoing debate in the literature on the extent and nature of "globalisation", such measure could be useful in assessing the nature of mergers and acquisitions particularly in areas which had been the subject of national industrial and technology support: if a

---

[1] See amongst others Narula, 1996, Pavitt and Patel, 1998.

firm, with a foreign headquarters, had a low degree of globalisation, the integration of a national firm into the former could be described in terms of it coming under foreign control. By contrast a national firm with a low degree of globalisation becoming part of a more global company, might provide it with new, global market opportunities.

But it will also be obvious from the description above that such measures are likely to become quickly eroded by the many practical ways in which such explicit expressions of globalisation can become faked or hidden. There is an obvious need for a more international policy response. The need for some international rules of the game, in particular with respect to competition policy arises precisely from national differences in competition policy and the absence of an international regime for overseeing transnational investment, acquisitions and mergers. At the risk of increasingly becoming a source of international conflict in the few areas of international harmonisation and institutional authority (such as trade policy and WTO), such international policy should aim at counter-acting the emergence of world wide cartels between global firms; reduce the divergence between national competition policies, and monitor more closely the degree and extent of globalisation of such firms.

At the same time, and maybe paradoxically, the multi-domestic firm also questions the relevance of national policy making from a regional, local perspective. As indicated above, multi-domestic firms will both take advantage and contribute to the emergence of locational infrastructural advantage. Of particular importance in this context is the infrastructure linked to the innovation system. It is this infrastructure which provides the major incentives for private investment in intangible resources, including human resources; for linking up with public research institutes (possibly assisting in setting up specialised centres of excellence, training partnerships, technical information agencies, etc.), that might lead to a local interactive learning cluster and possibly even the establishment by such firms of a global so-called "competence centre" in a particular product or market niche.

The effective exploitation as well as the contribution of multi-domestic firms to such locally created advantages raises again a number of important policy issues. At the site level this might often translate into rivalries concerning the services offered to firms and there will in effect be no limit to the bidding. As a result, there is, as is evident from European experience, a multiplicity of new growth sites, science parks or technopoli, being set up; none developing the necessary size to reach some of the essential externalities and increasing returns growth features and all increasing the cost of communicating and interacting.

The desire of local authorities to attract at the local level such high tech learning centres illustrates to some extent the further erosion and relevance of national policy making in this area. Nowhere is this becoming more obvious than in cross-border peripheral regions where the national central interests are unlikely to coincide with local interest. In my own region (South-Limburg in The Netherlands) e.g., national policy and priority setting e.g. with respect to infrastructure or foreign investment attraction, is increasingly perceived as a form of "randstaddemocratie".[2] While the intensification of global competition has made the role of regional conditions, including regional policy, more important, the individual citizen is increasingly identifying such local conditions - - the quality of the environment, children's education, availability of social and cultural services - as the essential features of his welfare and quality of life. Hence the growing political pressure for decentralisation or devolution of policy responsibilities, including the necessary financial means, away from the national centre towards local communities (regions, cities, etc.). With the erosion in national government responsibilities, citizens themselves appear increasingly to request that a larger part of their national tax payments contribute directly to the improvement of their local living conditions. The effectiveness of such policies can then also be assessed in a much more direct and immediate way.

3. National Technology Support Policies and International Competitiveness

From a national policy perspective, economic and social progress can be said to depend on widespread capacity to compete in increasingly global markets and the dynamic turnover of winners and losers as efficiency in exploiting new economic opportunity shifts between enterprises and nations. This has been to some extent the bread and butter of national economic policy. The competitiveness question is whether technology is today such an important element in the process of structural change and globalisation that differences in the capacity to bring technology into the market are a matter of priority concern for enterprises and governments? And whether it is simply a matter of enterprise strategies and capabilities, or whether public authorities need to intervene to ensure that their enterprises can compete in the international market?

The old debate about different North American, European and Asian capabilities can be seen in this light. It is not so much an issue of access to technology but to the capacity to innovate and to diffuse technology. These capacities depend then on a wide range of conditions and institutions, some of which might be strongly influenced by government policy, although the essential feature of success is likely to be entrepreneurship, involving technology, management and financial innovation.

Given the great variety of institutional set-ups, can one identify some regularities across industries and across countries? In order to provide some tentative answers, it is essential, as has been pointed out by many economists in the (neo-)Schumpeterian tradition (from Dosi, 1984 to Howitt, 1996) to distinguish between "normal" technical

---

2 A term used by the chairman of the local Chamber of Commerce.

progress which proceeds along the trajectories defined by an established paradigm and those "extraordinary" technological advances which relate to the emergence of radically new paradigms.

As regards the latter, it is generally acknowledged that market processes are generally rather weak in directing the emergence and selection of these radical technological discontinuities. When the process of innovation is highly exploratory, its direct responsiveness to economic signals is rather loose and the linkages with strictly scientific knowledge have been quite strong. In such case non-market organisations appear to have played an important role, providing often the necessary conditions for new scientific developments and performing as ex-ante selectors of the explored technological paradigm within a much wider set of potential ones. One may remember in this context the case of the semiconductor and computer technologies and the influence of both military agencies and big electrical corporations in the early days of the development of these radically new technologies. Somewhat similar cases can be found in the early developments of synthetic chemistry,[3] or more recently the development of bioengineering, new materials or even the Internet.

In general, the features of the process of search and selection of new technological paradigms are such that the institutional and scientific context and existing public policy are fundamental in so far as they affect (a) the bridging mechanisms between pure science and technological developments, (b) the criteria and capabilities of search by the economic agents, and (c) the constraints, incentives and uncertainty facing would-be innovators.

The counterpart of this proposition at the international level is that, when new technologies emerge, the relative success of the various countries or regions in the world will depend on the successful matching between each country's scientific context and technological capabilities; the nature of their "bridging institutions"; economic conditions (relative prices, nature and size of the markets, availability/scarcity of raw materials, etc.); and the nature of the dominant rules of behaviour, strategies and forms of organisation of the economic actors. All these variables are also but to different degrees affected by public policies, either directly (e.g. procurement policies or R & D subsidies which obviously influence the economic signals facing individual firms), or indirectly (e.g. through the influence of the educational system upon scientific and technological capabilities, the effect of taxation policies on the emergence of new firms, etc.).

As regards "normal" technical progress, the variety in the organisational patterns of innovative activities is of course much greater and makes it difficult to draw some general trends. Two have been highlighted in the literature.

---

[3] For more details see Freeman and Soete, 1997.

First, there is a technology and country specificity of the balance between what is co-ordinated and organised through the visible hand of corporate structures and what is left to the invisible hand of the markets (Pavitt, 1984, Tidd et al., 1997). In science-based industries for instance, whenever technological paradigms become established, the process of Schumpeterian competition tends to produce rather big oligopolies which also internalise considerable innovative capabilities (e.g. computers, semiconductors, synthetic chemicals, software, content, etc.). In production intensive industries in somewhat similar fashion, the "visible hand" of big corporations puts the organisation of technological advances at the core of their strategic behaviour (e.g. automobiles, most other consumer durables, etc.). In the case of specialised suppliers, technological advances are generally organised through the matching between the own specific technological skills and intense (often arm-length and untraded) relationships with users or component producers. Finally, only in supplier dominated sectors do the mechanisms of organisation and co-ordination of technological progress appear to retain some significant similarities with the classical view of the "invisible hand": technological advances are generally available on the market in the form of new capital goods, there are many firms generally with weak strategic interactions, etc.

Second, there are significant intersectoral differences in the balance between public institutions and private organisations in the process of innovation (Mansfield, 1995). Some sectors rely on an endogenous process of technological advance while others depend heavily on public sources of innovation. In Dosi, Pavitt and Soete (1990), we suggested the following empirical generalisation: the higher the role of the visible hand of oligopolistic organisations, the lower the requirement for strictly public institutions in the processes of economic co-ordination and technological advance. And vice versa: the nearer one activity is to "pure competition", the higher its need for strictly institutional forms of organisation of its "externalities" and technological advances. Agriculture is a well-known case in point: historically, a significant part of its technological advance, at least in the US, has been provided by government sponsored research. By contrast, many oligopoly dominated manufacturing sectors have produced endogenously a good part of their "normal" technological advance, and have appeared to coordinate their price/quantity adjustments rather well.

The foregoing discussion suggests in other words that in the post-war economic development non-market agencies on the one hand have been a major actor in the emergence of new technological paradigms, whereas on the other hand the conditions of technological opportunity and appropriability have been such as to guarantee rather sustained rates of "normal" technical progress endogenously generated through manufacturing oligopolistic corporations. It is important to note though that every government, has intervened, in forms and degrees that depended on the sectors and countries, so as to strengthen the incentives to innovate.

This is also the case with respect to the European economic integration process, discussed in the next section. Until the early 1980s, science and technology policy in Europe was dominated by national programmes. The role of the European Commission in R&D was limited to nuclear research supported by the Euratom Treaty of 1957. After the early 1980s various research policies were developed and again given civil legal support by Title VI of the 1987 Single European Act. The focus of most of these research policies, developed, implemented and monitored by the European Commission was, as in other areas, primarily directed towards programmes to overcome the fragmented, national structure of European industry and markets. One of the main purposes of these policies was to permit economies of scale. However, no European "national system of innovation" emerged as yet (Caracostas and Soete, 1996); on the contrary the European institutions were often simply added to existing national and regional institutions and instruments.

Confronted with this variety of organisations, degrees and forms of public intervention, can one make any normative statement linking institutional forms, degrees of public involvement and economic performance which might be of relevance to a discussion of future growth and developments paths particularly within the framework of the European integration process?

In a rapidly changing complex world with not just a gradual planned process of European integration but also of increased globalisation, one can hardly reach definite conclusions on "optimal" set-ups. Leaving for the moment the particular issue of European integration for a separate discussion in the next section, one can, at best, define some trade-offs involved in each national organisational configuration. A good example of such a trade-off analysis was the "Challenging Neighbours: Rethinking German and Dutch Economic Institutions" project of the Netherlands Bureau for Economic Policy Analysis, a project started a couple of years ago from the perspective that Dutch institutions could learn a lot from what were perceived as "best" or at least "better" practice German institutions, but once finished providing more the opposite inside to policy makers in both countries.

Within the context of technology policy three such trade-offs appear essential. First, at the very essence of the innovative process undertaken by profit motivated agents there is necessarily some sort of "market failure" in a static sense. Varying degrees of appropriability are the necessary incentive to innovate, but imply at the same time "excess profits" and "sub-optimal" allocation of resources. Best practice techniques and better products diffuse throughout the (national and international) economy after a lapse of time and the gap between the technological frontier and the infra-marginal techniques also measures to some extent the static inefficiency of any pattern of allocation of resources.

The asymmetries in capabilities are a direct consequence of the partly appropriable nature of technological advances. This feature also corresponds to an asymmetric pattern of economic signals so that high technological opportunity, associated with a high degree of appropriability of technological innovation may well perform as a powerful incentive to innovate for a company which is at or near the technological frontier. At the same time such technological opportunities will be a powerful negative signal (an entry barrier) for a company with relatively lower technological capability. The current development of the software industry and its geographical concentration in the US (Steinmueller, 1996) following the increasingly successful enforcement of intellectual property world wide appears a good case in point.

A second normative issue concerns the ways each society builds its technological capabilities and appears to have translated these into innovative entrepreneurial behaviour. Again, one observes a wide international variance in both the "supply of entrepreneurship" and the ways it is institutionally formed. The difference between the "organised entrepreneurship" of Japanese firms and the self-made-man archetype in the US is a typical example; or between the formalised "production" of technological/managerial capabilities in France (the Ecole Polytechnique, etc.) and the anarchic Italian pattern. Many historians have provided suggestive descriptions of the growth of American technocracy, which highlighted the enormous changes which the contemporary economies underwent since the times of the "classical" protestant capitalist studied by Weber in <u>Protestant Ethic and the Spirit of Capitalism</u>. Yet more international studies on the mechanisms of formation of managers/ technocrats/ entrepreneurs would be needed in order to understand the social supply in the various countries of this crucial factor in innovative activities. The EU policy call for more entrepreneurship (one of the recommendations of the Jobs summit in Luxembourg, EU 1997) should be understood in this context.

A third normative issue concerns the possible trade-off between allocative efficiency and flexibility, or, more generally speaking, between "fitness" to a particular state-of-the-world and capability of coping with other (and unpredictable) environments. One can detect here an analogy with biological evolution. Extreme optimisation within a given environment might well imply a "dinosaur syndrome" and inflexibility to change. Conversely, high adaptability is likely to involve waste, "slack" and sub-optimal use of resources.

There is little doubt that the current diffusion of new information and communication technologies has substantially shifted the trade-offs between flexibility and economies of scale, increasing flexibility, lowering the minimum throughputs which allow for automated processes and shortening product life cycles. There is today a much more significant requirement for variety in capabilities, behavioural rules, and allocative processes which might allow for greater adaptability to uncertainty and change. One of the greatest strengths of capitalism has been its capability of continuously producing redundant resources, of exploring an "excessive" number of technological trajectories, of producing a wasteful number of technological/organisational "genotypes". In a sense and contrary to the old general equilibrium notion, if there is some advantage of contemporary market economies as compared to centrally-planned ones, it is probably the fact that the former do precisely not achieve an equilibrium of the Arrow-Debreu kind but are highly imperfect and always characterised by allocative inefficiencies and technological slacks.

The policy questions are consequently - and not surprisingly - rather complex. How can a sufficient "variety" be continuously generated? On the other hand, how can the potential of new technologies be better realised? To what extent is the realisation of such potential primarily depended on individual entrepreneurship and risk taking? Is the current move towards a more stringent and world wide enforceable appropriation regime (in patent law, copyrights, authors rights) slowing down international technology diffusion and raising technology related monopoly rents? These issues become even more entangled within the framework of the slow, painstaking economic integration process in Europe within an increasingly global "planet" environment of the 21st Century.

4.   European Economic Integration: from a Single Market to Diversified Technological Capabilities

Following the analysis presented in Soete (1997), the characteristics of past European economic integration can be summarised along the following three lines.[4]

First and foremost, economic integration has been inspired by the obvious desire to reap the scale advantages of a large, "harmonised" internal market. In manufacturing, this process of intra-European integration has more or less come to an end. Much of the European growth and employment boom of the late 1980's, as well as the wave of foreign direct investment (FDI) inflow into the EU, can be directly associated with the expected growth opportunities of the then forthcoming Single Market. Since then, and somewhat paradoxically in terms of the 1992 timing of the formal Single Market creation process, extra-European pressures for restructuring in manufacturing have taken over and increased rapidly, e.g. through the opening-up of Eastern Europe and the rapid export-led growth industrialisation pattern of many Asian economies.

In services by contrast, the intra-European economic integration process is still in its very first initial stages. The much awaited upcoming liberalisation of the telecommunications sector across most member countries will be the first clear case of the opening-up of a major service facility. Most other service sectors (public utilities, transport) are still relatively closed economic sectors. The difficulties and slowness in opening-up

---

4   In contrast to most current debates on economic integration I do not address the issue here of monetary union.

such service sectors within the EU contrast sharply with the ease and speed of the international opening up to international trade and competition in the WTO and in many of the new entrants. While the Commission as an institution is still playing a major role in such world-wide trade liberalisation discussions, the extra-EU pressures for rapid liberalisation and world wide integration are in the process of taking over the carefully planned but slow intra-European liberalisation and integration process.

An interesting question which, in my view at least, has not received enough attention in the economic literature is the trade diversion versus trade creation impact of Europe's economic integration process as it has taken place over the last two decades. An interesting hypothesis, which I already suggested a couple of years ago when analysing the poor performance of the European electronics industry, is that trade diversion has indeed dominated some of the most technology-intensive sectors. European firms as well as the subsidiaries of foreign firms have been "diverted" towards the easy European member countries' markets, and have foregone the, from a competitive and new product point of view, tougher US and Japanese markets. The result has been an increasingly poor performance in non-EU markets in some of the most dynamic, growing sectors. The wave of foreign direct investment in the various EU-member countries, starting already in the 60s and 70s, and accelerating in the 80s in view of the forthcoming "Single Market", has generally been of the "tariff-jumping" kind, aiming at presence in the world's largest consumer market and hoping to reap the benefits of such harmonised internal market, did in effect amount to some kind of import substitution industrialisation growth process. In doing so they (the US, Japan) could simply transfer to Europe the core competence and knowledge acquired at home of producing for large standardised markets.

From this perspective the actual economic integration process as it proceeded in Europe could well be compared with a gradual, unwarranted import substitution industrialisation growth process whereby the overall extra European competitiveness particularly in high tech sectors became gradually undermined. It is what could be called the "fortress paradox" of European integration: as Europe thought it would become better able to defend itself through the creation of its own large internal market, it became weaker because it left the most dynamic external markets to its competitors.

Second, to offset the possible negative effects of increased specialisation on trends towards uneven growth and regional divergence -- something many so-called new trade economists have been pointing to --, the European economic integration process has been accompanied by a clear policy of financial transfer from rich to poor countries. Hence, "cohesion" became the major second policy aim and was expressed through the creation of European Structural and Social Funds that aimed at developing better infrastructural provisions in peripheral and less favoured regions. In some of these countries/regions such funds became the most important source of public investment.

In prioritising "cohesion", the European economic union became gradually characterised by an economically integrated zone with free movement of goods, consumers and financial flows, but not of labour. Rather the contrary, despite the desire to also achieve the free movement of labour, the extent of intra-European migration declined with each new enlargement of the union. While such limited intra-European labour migration fit the objectives of European cohesion, i.e. to transfer financial resources to less favoured regions and create employment opportunities rather than have employment migrate to richer regions, the lack of intra-European migration reduced in a significant way possible adjustments in the labour market at the European level, and in particular possible adjustments to shifts in structural change, such as globalisation. Only in a limited number of high skilled areas did mobility increase in any significant way, reinforcing rather than reducing intra-European growth divergence.

It is what could be called the "migration paradox" of European integration: as goods and capital flows became more mobile across Europe, labour became more immobile, further segmenting labour markets at the national level.

Third, the economic integration process was accompanied by a set of specific European industrial and technological policies, fostering intra-European co-operation in the field of pre-competitive R&D, university researchers, students, and various support programmes for particular technology fields: the so-called framework programmes and other related technological support programmes. Interestingly, these policies that aimed at strengthening European competitiveness in high tech sectors have probably been most successful in some of the "big science" RTD areas, where essential scale economies could indeed be achieved. In most other areas though, the EU resources available when compared to national resources were too limited to make any impact in shifting or redirecting countries' own national priorities in supporting investment in knowledge accumulation (both education, training and research). At the same time, the international accessibility to codified knowledge increased dramatically through the use of ICTs. While support for intra-European research collaboration might still be welcome in many cases, the essential research collaboration will often be of a much more global nature, going well beyond the European borders. Here too, there could be a case of knowledge acquisition "diversion", the intra-European exchange having taken place at the expense of extra-European exchange. In the more basic research areas where open international access has always existed, such "diversion" might have ultimately had little impact; in the more applied business research areas, it might well have been one of the factors behind the dramatic growth in so-called "strategic alliances" between large European, US and Japanese firms trying to source knowledge more globally while at the same time benefiting from various national or supra-national support programmes.

It is what could be called the "European paradox": as Europe invested in intra-European research, in the collaboration and exchange of scientific knowledge among European scientists, or even in the technological strengthening of the competitive potential of European firms, the advantages of such geographically "bounded" collaboration have become marginal, given the dramatically increased opportunities for the fast exchange of information and co-operation.

In listing these, for the unwarned reader, somewhat peculiar characteristics of Europe's economic integration process, I realise of course that I have painted a rather one-sided picture of what I consider to have been some negative side effects of the process of economic integration as it has taken place in Europe over the last ten to twenty years. My main point will hopefully be clear: the "diversion" effects accompanying intense integration processes such as the forming of the European Union, can take many forms. In the case of Europe, the simple fact that this integration process was accompanied by a much faster "external" world economic integration process might well have led to a systematic diversion away from some of the most significant new trade opportunities linked to globalisation.

5.   Conclusions

The new challenges brought about by globalisation imply to some extent the need for policies which focus more on the peculiar characteristics of the enormous variety in European development, and try to build upon these to develop new dynamic growth opportunities. It means in the first instance acknowledging that the reaping of industrial scale advantages and the need for regulatory harmonisation which have characterised European economic integration so far have to some extent reached their natural limits and can be further pursued within the broader world economic context. In a more general sense it also means recognising that there has been an over-preoccupation in Europe with labour efficiency improvements and process-oriented technological change, reflected e.g. at the macro-level in a systematically lower capital labour substitution elasticity than in the US or Japan. While there is little doubt that the achievement of scale advantages will continue to be one of the major challenges in many new sectors, such as new information services and products heavily dependent on scale economies, there is also little doubt that European competitiveness and extra-European growth opportunities will have to depend on something more, something specific to Europe.

Indeed, the economies of scale in many information goods are often even more dramatic and significant than in the case of manufactured goods. The lack of a harmonised European market in many basic services sectors is a major cost factor and has undoubtedly an overall negative impact on European competitiveness in many other sectors. In information services the fragmented European market is undoubtedly a major barrier not just for the rapid diffusion of information services but also for the emergence of a competitive European multi-media industry. But even in this case it will be obvious that policies which would simply aim at reaping the advantages of scale economies would in the end undermine some of the essence itself of European competitiveness based on its widespread cultural, educational and social diversity. The guiding policy principle can to some extent no longer be that the EU contains one of the worlds largest consumer markets of 350 million, but that the EU contains one of the most culturally, educationally and socially diverse markets with, as David Putnam put it "a potential of 350 million producers". From this perspective, the current world economic integration process signals the need for Europe to develop a new, different economic integration process. This process no longer puts the sole emphasis on the need for the standardisation and harmonisation of products and services, access to "open" infrastructure, and improved transparency of markets across Europe. Instead it recognises and nurtures the many differences in tastes, cultures and talents.

The extent to which such new policies, reflecting in many ways the desire for more decentralised, nearer to citizen decision making both in business and government, can indeed enhance this "productive" potential of Europe's enormous variety into competitive advantage is likely to become the central question that will have to be addressed in the coming years. It relates to the degree to which the size advantage of the more than 350 million inhabitants is not only translated into the satisfaction of common material and information needs at lower prices, but also into a productive creativity potential and communication and exchange needs of diversity and variety. It is in this sense that location of production does indeed matter, even in a world which increasingly looks like a village.

References

Dosi, G. (1984), *Technical Change and Industrial Transformation*, London, Macmillan.

Dosi, G., Pavitt, K. and Soete, L. (1990), *The Economics of Technical Change and International Trade*, Brighton, Wheatsheaf.

Freeman C. (1987), *Technology Policy and Economic Performance: Lessons from Japan*, London, Pinter

Freeman C. and Soete, L. (1997), *The Economics of Industrial Innovation*, Third Edition, Creative Print and Design, London.

Howitt, P. (ed.), (1996), *The implications of knowledge-based growth for micro-economic policies*, Industry Canada Research Series. Calgary: University of Calgary Press.

Lundvall, B.A. (ed), (1992), National Systems of Innovaton: Towards a Theory of Innovation and Interactive Learning, London, Pinter.

Mansfield, E. (1995), *Innovation, Technology and the Economy, Selected Essays*, 2 vols. Aldershot, Elgar.

Narula, R. (1996), *Multinational Investment and Economic Structure*, London, Routledge.

Nelson, R. (1992), What is "commercial" and what is "public" about technology, and what should be?, in Rosenberg, N., Landau, R., Mowery, D.C. (eds.), *Technology and the wealth of nations*. Stanford: Stanford University Press.

Patel P. and Pavitt K. (1998), Uneven (and divergent) technological accumulation among advanced countries: evidence and a framework of explanation in D. Archibugi and J. Michi (eds.), *Trade, Growth and Technical Change*, Cambridge: Cambridge University Press, 1998, pp. 55-82.

Pavitt, K. (1984), Patterns of technical change: towards a taxonomy and a theory, *Research Policy*, vol.13, no. 6, pp. 343-73.

Soete, L. (1997), The Impact of Globalization on European Economic Integration, *The ITPS Report*, no. 15, pp. 21-28.

Steinmueller, E. (1996), The U.S. software industry: An analysis and interpretive history, in D.C. Mowery (ed.), *The International Computer Software Industry: A Comparative Study of Industry Evolution and Structure*, Oxford University Press, pp. 15-52.

Tidd et al. (1997), *Managing Innovation: Integrating Technological, Market and Organizatinal Change*, Chichester, Wiley.

New Challenges to Education and Research in a Global Economy

Jürgen Mittelstrass

In a world that in economic affairs is oriented by the keyword 'globalisation', not only are economic conditions themselves undergoing changes, but also the conditions for education and research, especially with regard to the connection between education and research, that is, the system of academic education. It is a legitimate question whether this system is educating adequately if local education has to satisfy global needs and if at the same time work, including scientifically distinguished work is becoming scarce and professional profiles are changing under the influence of economic and technological developments.

The usual keyword in this connection is *practical orientation* and includes as a rule, at least in Germany, the reproach that the academic system has taken too little heed of just this keyword in its educational structures. According to a stubbornly assertive prejudice, this system still displays the architecture of ivory towers and produces for itself and for our society a new generation that knows a great deal and can do very little, that is inadequately trained for the working world, especially for one in a process of rapid change, and that is not ready for the tough climate of a professional system which places strong demands not only on the quality of education but also on the transformation of knowledge into ability, that is, on the conversion of 'theoretical' knowledge to 'practical' performance. Indeed *theory* and *practice* are still - though in a much less ideological sense than in the socio-political debates of the 1960's and 70's - the key concepts which orient the current discussion about the proper academic educational system and conducive conditions for technological innovation.

It is possible, by the way, that in this form itself a prejudice is passed on, namely the notion that the academic system itself, and thus its educational forms, is theory not practice. As if learning, too, were not something extremely practical, that is, practice. In other words - and this is intended as an early warning against conceptual sloppiness: The boundaries between theory and practice do not run between textbook and profession, but rather arise in all forms of work, both in academic forms (including academic educational forms) and in professional forms. It is these working forms that are designated by the title "new challenges to education and research in a global economy". Under this heading let me make three points with some brief elucidations about globalisation, research, and education.

> **1. In a global economy not only do economic structures change but social structures as well. Institutional isolation dissolves. This development affects both education and research.**

There are some words that saunter up on velvet paws and then whip out their claws. Globalisation is one such word. It means, as we all know, the free transfer of raw materials, commodities, capital, services, and labor across all geographic and political boundaries. The concept of *internationalisation*, on the other hand, indicates a growing proportion of international trade and its increasing interlocking as well as the motions of capital, labor, and know-how between different national economies and their economic agents - and in this sense is a concept derived from national relations. In contrast, such limitations disappear in the case of globalisation. What is global is not derivative, but what is given first; economic and political boundaries, which thus far have determined at least the 'beginning' of economic action, are dissolving.

Furthermore, the concept of globalisation is not merely an economic concept or one of economic policy, to the extent that these concepts refer only to economic activities from the perspective of competition. The truth is "that globalisation comprehends more factors than were observable in earlier stages of development and that our entire social and institutional fabric will change fundamentally. Even if globalisation (...) is economically induced, the consequences extend far beyond this area and have thus far been little understood - especially in their significance for us in our social relations and organisational structures."[1] In this sense the concept of globalisation comprehends the new information and communication technologies as well, which are largely not subject to local control, the rise of supra-national political institutions (keyword: 'globalisation of the political'), and the increasing homogenisation of education and research structures. Not only is an economic and political dimension defined but also general social and cultural dimensions, in as much as globalisation "harmonises consumption patterns, labor organisation, and institutional preconditions, creates new polarities between highly efficient globally operating enterprises and local 'backward' organisations (including state administrations) and occupational groups, induces ecological long-range effects through growth dynamics and the spread of 'non-enduring' consumption patterns, evades traditional control instruments of the social balance of power (from cartel legislation to the jurisdiction of national parliaments), further reduces the predictability and planability of developments and continues to accelerate technological development."[2]

---

[1] Cf. Steger (1996), p. 4.

[2] Cf. Steger (1996), p. 5.

If this analysis is right - and there is little doubt that it is - then globalisation is the keyword for a multidimensional transformation process of modern society into a - rather imprecisely determined - future. Not only are enterprises in the traditional sense being dissolved by becoming 'virtual' enterprises, that is, by being replaced by a network of regional independencies, but the same holds for social structures, which up to now have been defined essentially by national and cultural stabilities. Among these are education and research.

> 2. Research up to now has mainly been defined by the distinction between basic research and applied research. This distinction no longer holds water, i.e., all research forms have a dynamic relationship to one another. They form a research triangle.

Science today is drawn into applications and developments, and this is not only due to the process of globalisation but also to inner-scientific developments. Whereas earlier science defined itself, in accordance with the ideas of truth and pure knowledge, almost exclusively through the concept of pure or basic research, things have become somewhat more complicated today, in spite of the retention of the old and cherished distinctions and evaluations. In particular, the old distinction between basic research and applied research, in which basic research was considered science and applied research more or less business, is ever less appropriate to the real situation of science at the level of development of modern societies under the impact of globalisation.

This does not mean that *pure basic research* does not exist anymore. It can be identified as such wherever research is being carried out, the results of which show no recognisable practical application. Typical research fields of this kind are high energy physics and cosmology, for instance, the development of a unified theory of all non-gravitational particle interactions or the connection of the theory of elementary particles with the theory of gravity. In these cases the notion of application makes no sense, not even a predictive sense (unless one wants to take up science fiction). This research is pure basic research. Somewhat different is the case of what we might name *application-oriented basic research*, a type of research from which we expect applications in the long range, but not of the kind that could be directly marketed or developed within the normal planning time spans of industrial enterprises. Examples of this would be high-temperature superconductivity, synergetics (non-linear thermodynamics) and the foundations of information sciences. In these cases application is intended even though the paths from research to application are unclear and are themselves in need of intensive research. A third type should be distinguished from these two, namely, what we might call *product-oriented research*. This is research which takes place either with a view to particular application or which promises such application in the near future. Examples would be materials research, environment research, medical research (for instance AIDS research). In such cases the paths between research and application are short and are constitutive parts of the research programs.

These three types of research are often mutually supportive in concrete research programs both in and outside the university; they interlock and intermingle when focusing on a problem. Application-oriented basic research, alongside product-oriented research, is becoming more and more the norm. This also means that the goals of science, in as much as these are expressed by such ideals as truth and knowledge, are becoming more and more joined to the goals of a world that is less inclined to admire than to apply the results of science. In fact, neither the Greek mind, to whom we owe the idea of science, nor the modern mind, that created the modern world, cautioned science in its cradle to stay away from application. Nonetheless, with growing closeness to application, the responsibility of science and the scientist also grows. In our world, governed by a global economy and its social and institutional consequences, it has become harder, not easier to be a good scientist. For a new concept of research that is neither empty (because without distinctions) nor ideological (because linked to an obsolete concept of basic research), this means that we are dealing with an (equilateral) *triangle of research*, whose corners are pure basic research, application-oriented basic research, and product-oriented research. My assertion is that this triangle is also the essential form of research under the perspective of globalisation and technological innovation.

> 3. When economy becomes global and research moves in a research triangle without any restrictions, (academic) education has to move out of its disciplinary boundaries. The future of research and learning, i.e. the future of (academic) education, is problem-driven transdisciplinarity.

If the archaic simplicity (sometimes simple-mindedness) in research affairs has nowadays become a complex interlocking of interdependent research orientations, (academic) education, too, has to change. Up to now this education has been strongly oriented towards fields and disciplines. It is often overlooked that most of the problems that research and a good education is supposed to help us solve do not do us the favour of defining themselves in terms of fields and disciplines. And these problems are precisely the ones that are particularly urgent. Examples are environment, energy and health. There is an asymmetry between the development of problems and the development of disciplines, and this asymmetry is growing to the extent that disciplinary development is increasingly determined by specialisation. There are problems "whose discipline we haven't found yet,"[3] and which - against a background of increasing particularisation and atomisation of fields of study and disciplines - perhaps we never will find. Therefore, the opposite path, the return to

---

3 Cf. Krüger (1987), p. 119.

larger disciplinary and interdisciplinary units, is also the more promising alternative, for instance, in the case of environmental problems. Ecological problems are complex problems, they can only be solved by the co-operation of many disciplines.

Interdisciplinarity, however, should not be conceived of only as a *repair measure* that is necessary when problems outgrow the limits of a discipline. On the contrary, interdisciplinarity, properly understood, serves to recover an ability to view things scientifically that also facilitates the recognition of problems and problematical trends before they appear, that is, before they become critical. What is involved is a *competition for (scientific) problems*. Disciplines enter into this competition - and even compete with the world, which has problems of its own. The main thing is to have such "future" problems. Whoever looks only at objects, as is often the case within a discipline, can easily overlook the fact that we do not just live in a world of objects, but also in a world of appropriation, needs, and an increasing loss of orientation. This, too, is an element of globalisation.

Where interdisciplinarity is synonymous not with the expansion of a system in which the number of problems one can deal with becomes smaller and the lack of creativity among academics greater, but rather with the expansion of our capacity to deal with existing problems and to anticipate future ones, then it is not enough to conceive of interdisciplinarity as a need for scientific organisation. Inter-disciplinarity, rather, has to begin at home, in one's own mind - as an ability to think 'laterally', to question what no one has questioned, to learn what is not known within one's own discipline. Whoever relies here solely on lofty organisational activity in science, has already squandered the prospects of interdisciplinarity to further science and (academic) education.

Interdisciplinarity, furthermore, does not move back and forth between the disciplines or hover above them like the Hegelian Absolute Spirit. On the contrary, within the context in which disciplines are historically constituted, interdisciplinarity helps to overcome the splintering of disciplines whenever these are in danger of losing their historical consciousness. Actually, this is what I call transdisciplinarity. One could also say that the last word in science, research and (academic) education is not interdisciplinarity but transdisciplinarity. And this is also true with respect to a growing global economy, to a new concept of research which fits the needs of a changing society under the impact of globalisation, and to the conditions of technological innovation which lie at the roots of economic and social progress.

So much for the three points and their brief elucidation. Returning to education and the educational system, this means: Globalisation in the form of a global economy presents increasing demands not only upon knowledge and ability but also on the environment of knowledge and ability. Globalisation does not at all mean simply to homogenise the economy, markets and culture but rather to deal with differing environments. Whoever wants to act on a global scale must be very familiar with the local situations. Therefore, education must prepare the educated for changing demands in almost all areas into which they are educated. We don't leave the haven of school to enter a world we are accustomed to. This holds also and especially for a world that is globalising itself. The research triangle and transdisciplinarity are answers to this situation.

References

Krüger, L.: "Einheit der Welt - Vielheit der Wissenschaft", in J. Kocka (ed.), *Interdisziplinarität. Praxis - Herausforderung - Ideologie*, Frankfurt 1987.

Steger, U. (ed.): *Globalisierung der Wirtschaft. Konsequenzen für Arbeit, Technik und Umwelt*, Berlin/ Heidelberg/ New York 1996.

3.   The Firm's Perspective

Improving Local Conditions for Innovation
The Scandinavian Perspective

Yrjö Neuvo

1. Creating an Innovative Culture

Making innovations is far from being easy. No one denies this. Making innovations for making innovations easier to do thus sounds like an impossible task! Still, in order for a company to excel, taking this task seriously is something worth the effort.

It is important to understand that innovations are needed at all levels of the organisation and in all kinds of businesses. Innovations should be found, in addition to products, in production, management practices, in quality issues, or in other words, in all corporate processes. Innovation consists of four basic elements: 1) knowledge, 2) hard work, 3) excitement and enthusiasm, and 4) knowing the goals. These elements need continuous nourishment for innovations to flourish.

True innovations seldom come out of the blue. Factors like market drivers and customer needs, technologies and capabilities and company's strategic intent give the right direction, but it is also vital that company culture and atmosphere support creative thinking and acts without prejudices. And when a correct amount of competitive spirit is added and enough cross-academic know-how is available, a winning package may have been formed. It has to be remembered though, that innovative culture can be maintained only if the top management supports it.

Continuous learning, respect for the individual, customer satisfaction and emphasis on achievement are essential for maintaining the innovative culture. (figure 1)

Rapid interaction, readiness for change and an ability to put ideas into practice are signs of a well-functioning, innovative environment. Combining these with multinational activities creates a challenge to any company. The question of how to support best skills in every individual in order for them to be part of effective teams, which again should work effectively with other teams and even across cultural borders, is something where better solutions should always be thought of.

Figure 1:   Prerequisites for a Good Innovation Environment

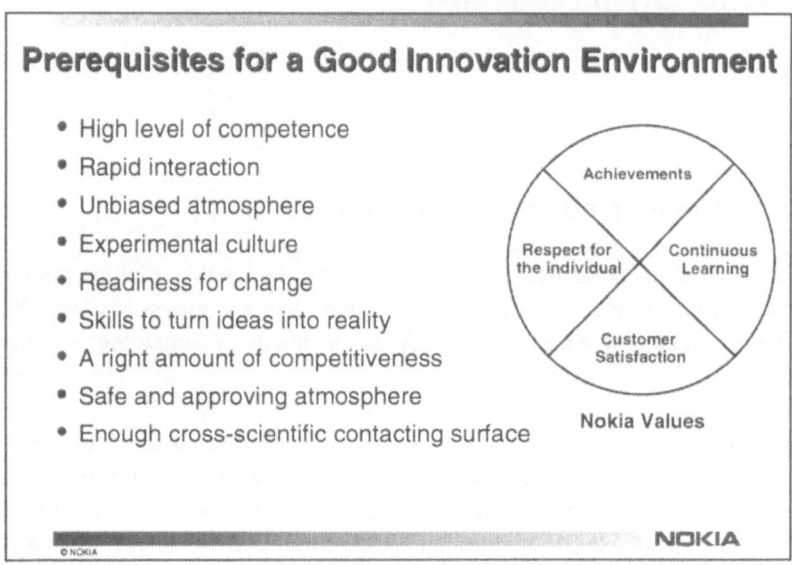

2.   Organisation and Work of Innovative Networks

In the telecommunications industry product life cycles are constantly reducing, and markets demand new products at an increasing pace. This causes pressure for the entire organisation, and especially for R&D, which needs to develop new products faster and faster, without making time-consuming mistakes. To achieve this, new ideas and new ways of doing things are required. Innovations are constantly needed.

Today many Finnish companies, including Nokia, have production as well as R&D units at several locations and in many countries. Having a multi-site R&D structure provides some potential advantages, but also some potential threats. Advantages sought with e.g. multi-site R&D may include better access to the sources of newest information from international research as well as better understanding of customer needs. This also allows easier access to qualified personnel in larger quantities. The challenges are in the creation of functional networks, division of tasks, and in pre-venting the undesirable side effects like the "not invented here" syndrome to pop up.

One way to overcome this kind of problems is to split the work into two well-defined phases. In the one phase communications are encouraged and in the other, the teams work in a more autonomous mode.

The networking mode is used to build up newest information which also is shared by all the R&D centres. This provides a homogeneous commonly agreed foundation for all the teams. This joint research-oriented activity implies strong team work across the network.

On this foundation the concrete development projects are then built with well-defined tasks for each of the teams. In this mode each team concentrates on its specific task and there is less communication in the network.

Here an essential prerequisite for successful networked activities is having a common language. Here a language has a broad meaning. It includes the way of operating, common terminology, and common goals. Building up this common language requires both time and definite actions.

3.   Co-operation between Industry, Universities and Government

The Finns have for some reason always been very keen to adopt new technologies. Recent proofs of this can be found e.g. from world's cellular phone penetration figures (figure 2). Finland holds the number one position with its almost 40 % penetration. As Sweden and Norway hold positions 2 and 3, it is no wonder two of the biggest players in telecommunications come from the Nordic Countries.

Figure 2:   Global Cellular Phone Penetration, End of September 1997

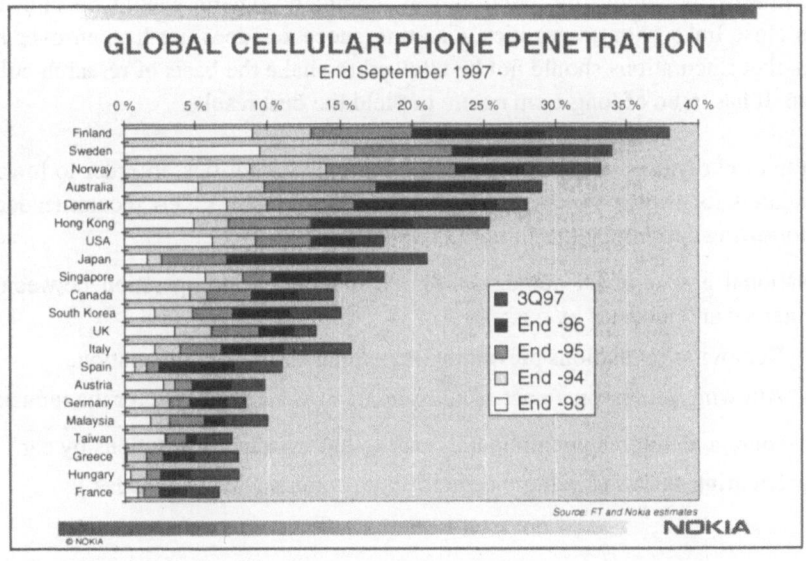

These figures are a good sign of how aware Finns are about the currently changing technological environment, and thus it is no surprise that we have a well-functioning and innovative high tech industry in Finland, including the very rapidly growing telecommunications business. But growth does not occur unless there are capable people feeding it, and we must make sure with our actions that elements for growth are available. For starters, this means the chain from the education system to industry works well, and government takes actions supporting this co-operation.

High tech requires a high level of competence. This is a product of good basic education and willingness to learn. Best results in gathering the most skilled and qualified people to specifically work in your company are often achieved by co-operation between industry and universities, as well as with national and regional authorities. In the beginning of 1997, the European Science and Technology Assembly, ESTA, published a report on 'Academic and Industrial Research Co-operation in Europe'. In this report, suggestions of how to enhance co-operation were proposed to all the three mentioned above.

Universities must not forget that their main mission is the cultivation of science and high quality teaching. However, this mission does not exclude things like contract re-search, active and effective communications with industry, and developing new ways of running day-to-day operations.

On the other hand, industry should respect the universities' mission and understand and apply best scientific methods and technologies in developing new products, production processes and environmental safeguards. Industry should also employ students, as well as provide research and specialisation opportunities for university researchers and scientists. In order to enhance an effective flow of information and thus nourish an innovative environment, company experts should be allowed to form close links with universities. Creating successes does not happen over night, so market fluctuations should not be allowed to shake the basis of research collaboration. It has to be of long-term nature to yield the best results.

On top of everything are the national and regional authorities. In order to lower the boundaries for fruitful co-operations between these three, ESTA recommended e.g. the following actions for the authorities to take:

- National and regional authorities should encourage co-operation between universities and industry by e.g.:
  - Removing regulations preventing beneficial forms of co-operation.
  - Allowing professors to act as consultants or work part-time for the industry.
- Encourage entrepreneurship in universities and research institutions by e.g.:
  - Securing access to venture capital for a promising business ideas.

– Rewarding universities actively supporting start-up companies founded by students and researchers.

• Encourage universities to identify focus areas of research in order to reach the highest international level of scientific achievements.

From the point of view of supporting innovations, especially the ones on this list regarding the encouragement of entrepreneurship are highly important. Good ideas die too easily, if there are no means to put them into practise. It is also a problem, admittedly that too often university education fails to provide students with even a basic understanding of entrepreneurship.

4.   CASE: The Oulu Science Park

One way of enhancing entrepreneurship, as well as co-operation between the three links in the chain mentioned above, are Science Parks and Technopoles.

The definition of a Science Park is:

"A Science Park is a location dedicated to knowledge-based organisations and where higher education, research and technology transfer occur.'

In Finland, the oldest Science Park is located in Oulu, a city around 200 km from the arctic circle (see figure 3). The Park consists of three elements interacting with each other: Oulu University, Oulu Technology Park and Oulu Centre of Expertise.

Figure 3:   The Oulu Science Park (Oulu Technopolis Ltd.)

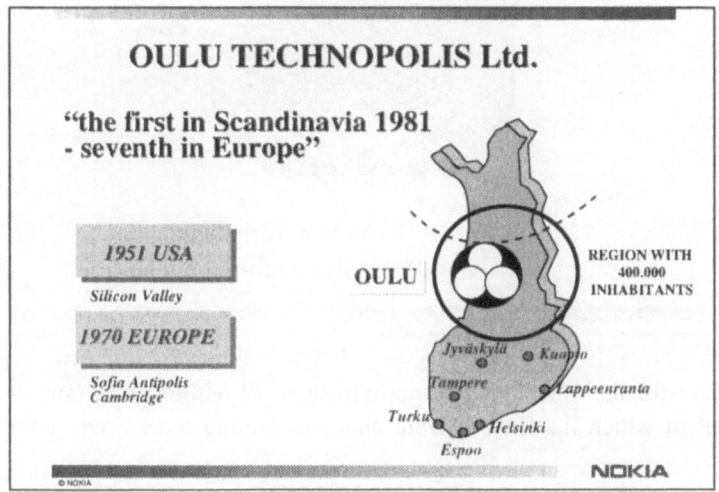

The Park was founded in 1981 as the first one of its kind in Scandinavia and seventh in Europe. It started with 18 companies employing 110 people, and today it is a dynamic business community consisting of over 100 growing companies, mainly operating in telecommunications, electronics and software. E.g. Nokia Mobile Phones' facilities in Oulu, for example, are located in the Science Park area.

The task of Oulu Science Park and its Business Incubator is to support its tenants and ensure the success of their high tech business. The tools used to achieve success include marketing support, development projects, good service and flexible arrangements regarding premises.

The Oulu Region Centre of Expertise has a significant role in developing high tech strategy in Northern Finland. The main values guiding its activities are:

- Emphasis on companies' needs
- Solidness of programs and actions
- Innovativeness
- Synergy
- World class level

Figure 4:   The Funding of Oulu Science Park

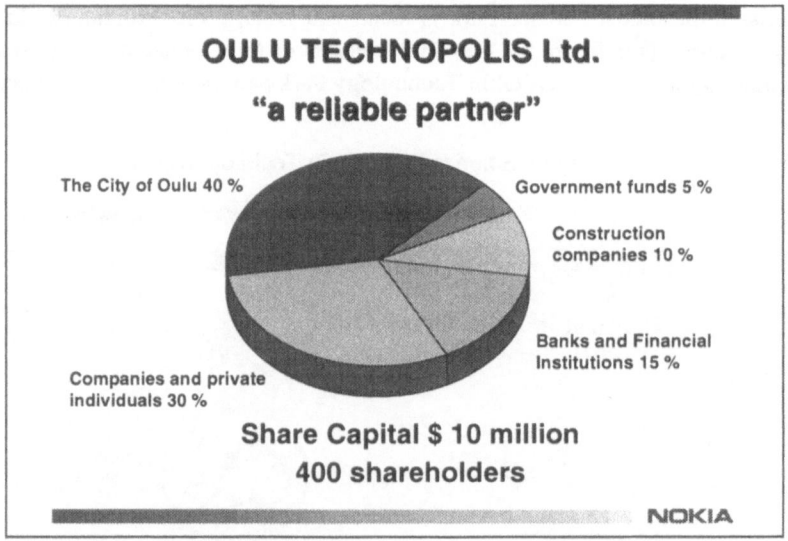

The budget for the year 1998 is approximately 12 Million Fim (app. 2,2 million USD) out of which the City of Oulu and surrounding areas cover with 4 million

Fim, while the rest comes from various sources like industry, EU, government etc. (see figure 4).

The Centre of Expertise program has a goal of creating 1,500 direct new high tech jobs in the area within the next years. Succeeding in this depends greatly on how well co-operation between education organisations and companies succeed and how effective company-oriented development projects are.

The Centre of Expertise program contains 10 subprojects during year 1998. Out of these, six will be closely linked and carried through by a group of high tech companies taking part in the Centre of Expertise program. The projects are called: Center for Wireless Communication (CWC), Mobile City, Pro Electronica, Columbus, Center for Wellness Technology (CWT) and Medipolis, From Idea to Products.

Practical knowledge about the Centre of Expertise has been gathered for several years. Here are some experiences from Oulu.

- The target of activities needs to be successful business
- Program needs to strive for world class level in its actions
- Clear and solid targets are the basis for success
- The time lag between an idea and a result is longer than expected
- Commitment of companies is a basic prerequisite for successful consortiums
- Companies will commit, if the contents of programs is in line with their long-range business development plans
- Role of local funding is crucial, and it requires national co-ordination and program
- Employment should be seen as a result of successful business

The results of the Oulu Region Centre of Expertise program have been good, and the program sees that the work has had a good effect on the area, and thus needs to be continued with more resources to become a national tool in developing high technology.

5. Conclusions

In order to be sure that the healthy growth of companies and economies is possible, innovations are needed in all levels and in all industrial functions. Good education is the corner stone for innovativeness and success, but universities, companies and governments must work together to achieve the best results. There must be enough courage and willingness for all of these parties to seek for new ways to lower barriers to more effective co-operation. Science Parks are a good example of good cooperative work. Business Incubators help promising ideas to airborne, major companies can benefit from the research done in the programs, and short geographical distances make communications easier.

Inside companies, a continuous effort should be made on all levels, and especially by the management, to create all the prerequisites for a good innovative environment. How to organise and manage R&D functions that are split around the world is a major challenge for any company, but when done properly, it will also turn out to be profitable.

In a well-managed innovative structure success breeds success and innovative teams increase their innovative power by them selves.

Global Development of R&D
The Japanese Perspective

Hans-Georg Junginger

1.   Introduction

This paper is focused to the Japanese perspective on global development of R&D. It will provide some insight as to how top management in Japan views the globalisation of R&D and what their expectations are for Sony's R&D efforts in Europe. Sony cannot be seen as a traditional Japanese company. Sony does not belong to a "Keiretsu". It is a relatively young company and just celebrated its 50th anniversary last year. If one looks at Sony's world-wide sales figures, the Japanese market generates only 28% of its sales. This is comparatively less than other Japanese multinationals. The U.S., including Canada, makes up some 30% of the sales, while Europe accounts for 22% and the rest of the world accounts for 20%. It is safe to say that Sony is among the most international companies in business today.

2.   The First Three Steps to a Globalised,
Multinational Company

How did Sony's globalization start? Sony has been single-minded from the start in its commitment to develop products that would appeal to the widest range of consumers possible, not only for its home market but also for international markets. "Global localisation" has been Sony's philosophy for that has driven and set the direction for expansion of its operations in international markets. However, like all Japanese electronics companies, Sony started in its home market and quickly moved to develop markets outside Japan by exporting to those markets of interest, namely the U.S. and Europe. Like in the U.S., early success in Europe led Sony to take a second step by setting up Sales and Marketing operations in different European countries. The first operation was set up in 1960 in Switzerland; and today, Sony has Sales and Marketing operations, including service operations, cover not only Europe, but also the entire Eurasian continent, from Lisbon to Almaty.

Sony took the next step by setting up its first local factory in Europe with the establishment of its colour TV facility in 1974 in Bridgend, UK. Though Sony's policy is to produce where the market is, political pressures, including the pressures created by "Fortress Europe", at the time gave Sony as well as other Japanese companies an additional incentive and push to begin setting up local assembly operations. Sony now has eleven factories in eight different European countries.

Figure 1:   Comparative Production - Europe vs. SE Asia

Now if today's production position of Europe is compared to that of South East Asia (cf. figure 1), one quickly recognises that things have changed considerably in the past years, fuelled by a rapidly changing and increasingly competitive market environment in Europe which demands greater market responsiveness than at any time in history.

The opening of Eastern Europe has also helped changing the equation. For example, products that are supplied by factories in South East Asia are on the ship for at least four weeks. In that time, prices not only fall, but market requirements also change, thus resulting in less competitive products. This means that the competitive position of Europe compared to South East Asia is strengthening. This, of course, varies according to product and industry. However, in Sony's case, European production for the local market offers a cost advantage, at the factory level, of four to seven percent on average compared to imports from South East Asia of the same product. Transport costs drop notably by producing locally; and if cost savings of 2% per month on average can be achieved by producing locally, then there is a strong case to do things locally. In fact, Europe has some of the most efficient Sony factories in the world (cf. figure 2). (Of course, currency exchange fluctuations can influence this picture considerably!)

151

The main cluster of factories in Europe and probably the largest factories are situated in Wales, in the UK, a result of the Thatcher effect that provided a lot of incentives to produce in Wales. The other factories are located in seven European countries. For example, the plant in Alsace produces GSM phones, while the plant in Barcelona produces colour TVs. And with Eastern Europe emerging as an important base, Sony has also started new factories in Hungary and Slovakia. In Hungary, the company makes VCRs and hi-fi products and in Slovakia TV components. Setting up factories in Eastern Europe was motivated by the opportunity to increase the overall European competitiveness in manufacturing. As these plants are still young, Sony is challenged to make these factories the most efficient and competitive ones.

3. The Fourth Step to a Globalised and Multinational Company: Factories

Building up efficient and competitive factories is only possible by localising engineering and component sourcing. It makes no sense to assemble locally and then import all the components; this brings no real advantage. In other words, local production requires local product development, or D&D (Design and Development), as Sony calls it. Products under development that are planned to enter production in the next three years would be treated as D&D at Sony.

Figure 2: Factories in Europe

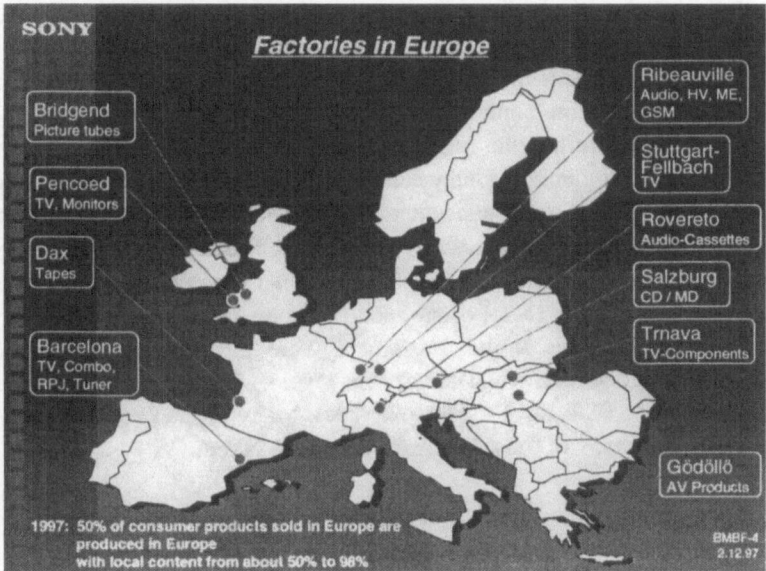

Figure 3:   Sony's European D&D Activities: Global Localisation

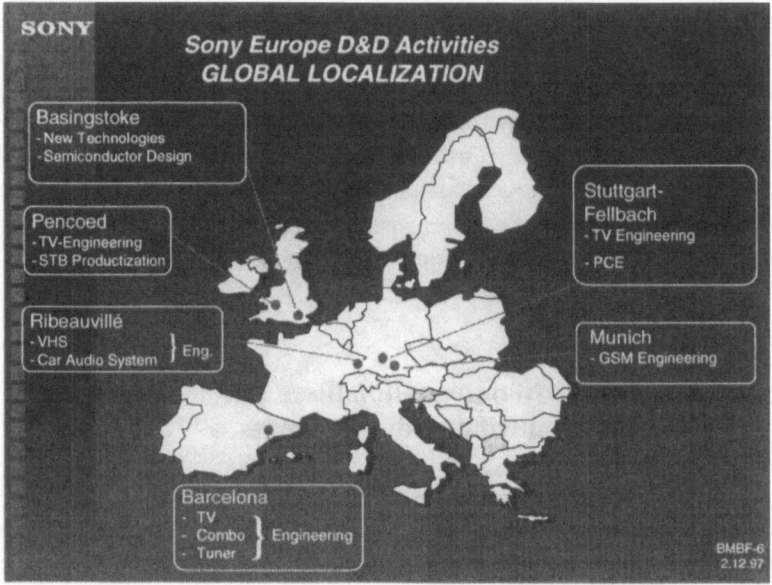

The company has localised D&D in the last years and has its main D&D centres in the UK, Germany, and in Spain (in Barcelona.). Local D&D combined with local sourcing of components has greatly contributed to Sony's ability and responsiveness to meet "European" market requirements (cf. figure 3). For example, Sony's TVs are among the most "European" sets on the market. Sony has TV engineering and product development on the continent in addition to its facility in the UK. This makes sense because the requirements to develop high quality TV sets on the continent, in Germany for example, are somewhat different to those in the UK. In other words, the company tries to reflect the different market needs in its product development activities. Today, Sony employs between 450 - 500 product development engineers in Europe. Half of all Sony's consumer products sold in Europe are produced in Europe. In comparison to its rivals, the company is ahead of its Japanese competitors and is at about the same level as the European competitors like Philips. The European firms also import a lot of products, like personal and portable audio products, from South East Asia. In fact, presumably Sony's production is more localised than Thomson's production. On average, our level of local content is at least 50% (but this goes up to 98% for TVs, for example). Many of the products have higher local content than many of those bearing a European brand.

4.   The Final Step to a Globalised, Multinational Company: Globalising R&D

What are the real challenges for the future, especially now that the world is undergoing a transformation from analogue to digital. For Sony one of the challenges is this final step of globalising research and development by establishing local (or regional) R&D centres. This is well underway at Sony, but the process will not be complete until the enterprise develops true global R&D centres, or as some call them, global competence centres. (For clarification's sake, R&D at Sony extends beyond a 3-year product development horizon, while D&D has a product development horizon of 3 years or less.) Sony's Top Management recognises that R&D will continue to play the most crucial role in Sony's future success, thus has been very active in pushing senior and middle management to continue efforts to strengthen R&D by developing competence centres outside Japan. Despite some resistance from the middle management in Japan, the globalization process of R&D is moving ahead one step at a time.

So what has been Sony's role and response in Europe to top management's challenge? To answer this, the European R&D mission is reviewed (cf. figure 4). Firstly, Sony undertakes R&D activities that support existing business in Europe. Secondly, the company undertakes activities to start and support new businesses based on European standards or by European initiatives.

Figure 4:   Mission - European R&D (I)

Since the arrival of European standards, like DVB or GSM, Sony's R&D investment and activities in this area have grown rapidly. Finally, the enterprise undertakes to use European specific know-how to support Sony corporate R&D and business group R&D in Japan and the U.S. Sony knows about the importance of adapting global R&D strategies and activities to reflect local market conditions, including regulatory frameworks and infrastructures. And now in a more complex world of digital technologies, the firm is forced to broaden its scope of R&D in order to develop the right solutions for a digital Europe.

Europe has more experience than Japan and the U.S. in pre-competitive research and co-operation. This is one of the reasons other companies are setting up R&D in Europe - to be part of such an environment where companies can easily co-operate with other companies in a pre-competitive way.

Based on these principles Sony is active mainly in its core business area of audio-video systems, as well as in the fields of telecom systems and what is called digital content distribution or interactive service systems (cf. figure 5). The enterprise believes in the future of technological convergence of computers, telecom, and consumer electronics; and thus is working increasingly in the areas of systems and standards. In addition, Sony started R&D last year in the areas of materials research, speech processing, and of course, environmental technologies, which is seen as part of its global corporate strategy.

Now, where does Sony locate its R&D activities in Europe (cf. figure 6)? For the most part, the company has two main centres in Europe, one in the UK where it conducts more broadcast and digital DVB based research that includes digital consumer R&D in the DVB area. Stuttgart (in Fellbach) is the second main centre and Sony focuses more on mobile systems as well as telecommunications. The enterprise is very active in the field of UMTS and DAB because the firm believes that DAB and UMTS and telecom activities are very strong in Germany. Sony also does materials research in Germany, and of course, environmental R&D. The company has also a very strong software team based in Brussels working on digital network solutions. There is no real reason to do this in Brussels. Sony's decision to locate there was the result of one person who was very effective in convincing top management to locate there. The company now has 60 software engineers in Brussels working on digital systems. Altogether in Europe Sony presently employs 230 engineers in R&D. Some years ago Sony employed 300 engineers in Europe, including both D&D and R&D. Now it has 700 and hopefully in the next three or four years it will probably double it again.

Figure 5:   Mission - European R&D (II)

Figure 6:   Sony's European R&D Activities: Global Localisation

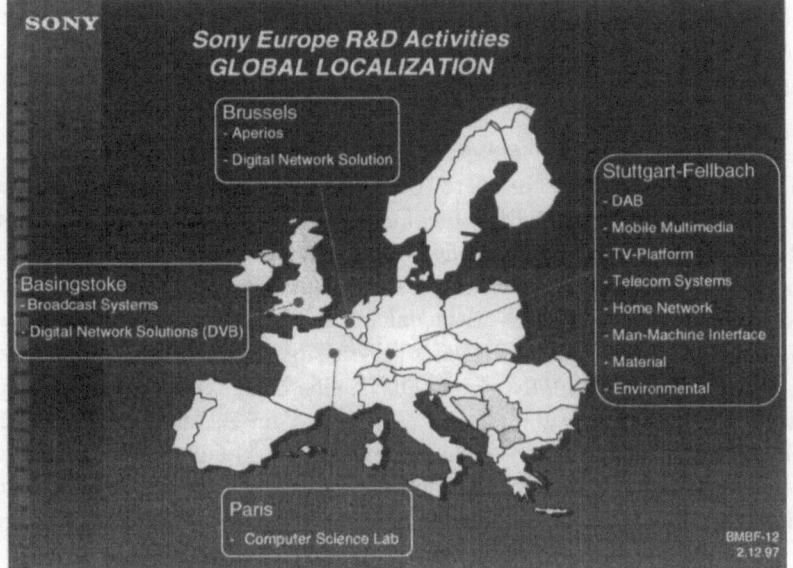

Figure 7:   New and Future R&D Area: The Digital Value Chain

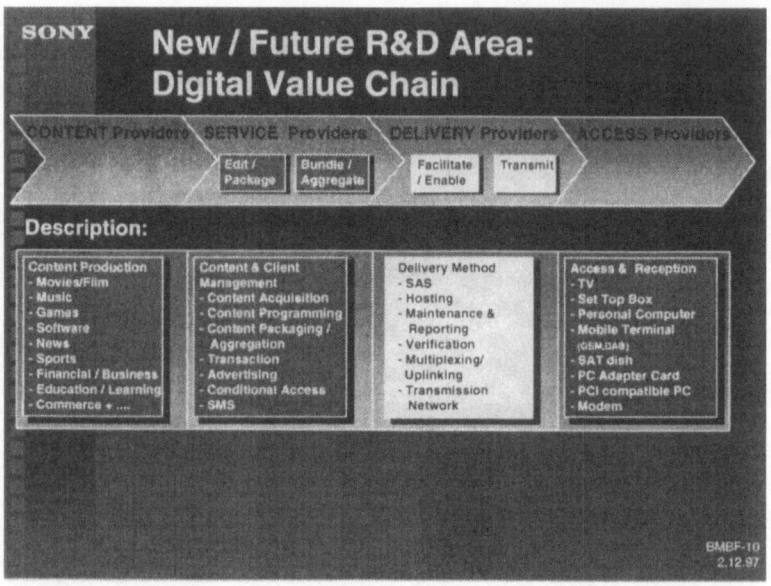

5.   New and Future R&D Area

A company like Sony will not only be active on the device or hardware side of the business in the future, but also will be increasingly active in all areas of the digital value chain (cf. figure 7).

Everybody understands the important role that content will play in the digital world and that Sony increasingly needs to focus on this area; of course, the company has come to know content as Hollywood and music, but also games. Sony cannot forget that games are one of the core applications driving developments in digital distribution and interactive systems. Sony is deeply involved on the content side; Sony Pictures (including Columbia Tristar) and Sony Music are fully owned Sony subsidiaries. The enterprise is now seeing very positive results from Sony's investment in Columbia Tristar. With the rise of Playstation Sony is very successful in games as well. And to create new high tech movies that, for example, could rival Jurassic Park requires significant investment in technology. The company is now working on such a project, Godzilla. Sony's content assets will only increase in value in the digital world.

Also in the digital value chain, the role of service provider (SP) is very important. The firm sees that in the digital value chain, Sony must also develop its presence in the business area of content packaging and aggregation. SPs in the digital world will be in a strong position to scope and structure the business as well as define the flow of revenue. SPs will use various transmission networks or delivery mechanisms to bring content and services to customers. There is a lot of work to do in this area - for example, in the areas of application programming interfaces (API), conditional access systems or subscriber management systems. The point is that the face of R&D is changing and that the company must prepare itself to meet this challenge, which can also be seen as an opportunity for Sony. It is here that the new ideas and revenues are being generated and new start-ups can be encouraged and created to help capitalise on this opportunity.

To say it in another way, Sony's traditional marketplace is changing and the company is quickly moving to a digital and networked world, one that is driven more by content/ services than hardware (cf. figure 8). In other words, customer relationships are evolving from a single contact with the customer at the point-of-sale, to long-term relationships with customers based not only on devices but also content and services. Sony is increasingly moving away from sales of stand-alone products based on separate technologies, to sales of products that are packaged with software and services based on integrated technologies. Finally, these changes require the enterprise to refocus and develop new R&D competencies. New start-ups should be encouraged and is a way to develop new competencies that would ultimately benefit larger companies as well.

Figure 8:   A Changing Marketplace for Sony

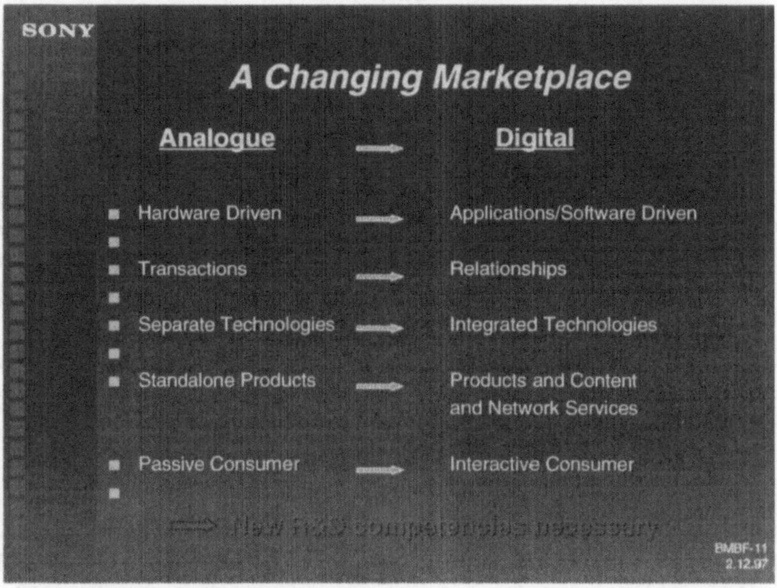

SONY

A Changing Marketplace

Analogue	→	Digital
■ Hardware Driven	→	Applications/Software Driven
■ Transactions	→	Relationships
■ Separate Technologies	→	Integrated Technologies
■ Standalone Products	→	Products and Content and Network Services
■ Passive Consumer	→	Interactive Consumer

→ New R&D competencies necessary

BMBF-11
2.12.97

Figure 9:   Expectations for R&D Environment

To illustrate this point, in Israel there are a lot of such activities going on by small start-up operations, where now a lot of the large companies are trying to get these competencies. Germany is facing a similar challenge.

6.   Expectations for R&D Environment

Now, what should be the expectations for Sony's R&D (cf. figure 9)? Though this whole area is still somewhat vague, it remains very clear, however, that if the company does something in R&D, it needs a skilled engineering base in Germany, mainly, of course, in digital technologies. Sony's know-how must not be limited to only hardware and software technologies. Sony, of course, does not want to see regulatory limitations placed on industry. It doesn't like the famous „Arbeitsstätten-verordnung" as far as it concerns R&D. It is also very helpful for the company to be offered incentives to work more closely with universities, research institutes and industry. The BMBF is working very successfully in this area.

On to more political issues, Sony should make Germany's strengths in R&D more transparent and actively promote them. For example, there are a lot of difficulties in Japan when making proposals to carry out certain projects in Germany. The Japanese think it makes no sense to pursue them because of the not so favourable image of Germany that is portrayed in the English media. So therefore, Sony has to bring the strengths of Germany to the forefront and avoid all this negative publicity. Finally, if all want to do something in Germany, everybody has to actively promote and strengthen the conditions compared with other European companies. The British are much more active, and of course, if they offer some incentives, Sony has to follow.

Improving Local Conditions for Innovation in the Pharmaceutical Industry

Norbert G. Riedel

This paper presents an overview of what it means to work in a global organisation, and how a global organisation affects the national industry in Germany. The case in point is Hoechst Marion Roussel, the pharmaceutical company of Hoechst.

It will be highlighted research as an example of a global effort, and that biotechnology has become a very major driver in research in general and in research in biomedical areas in particular. Furthermore this paper will describe how Germany can participate in what is a new wave in biotechnology that holds very great promise for employment as well as medical advances.

So the presentation really will touch on a virtual organisation to access innovation globally. Several topics are selected as they relate to the theme of this conference, but also as they relate to the philosophy and vision of Hoechst Marion Roussel as one of the major pharmaceutical companies in the world (cf. figure 1). One topic is innovation, since a commitment to innovation is absolutely critical in the pharmaceutical industry. The major players and the successful players in this industry will only be those who have innovative therapies in areas of unmet medical need, i.e. in diseases for which there is currently no sufficient or adequate treatment.

Figure 1:   Paradigms of the Philosophy of Hoechst Marion Roussel

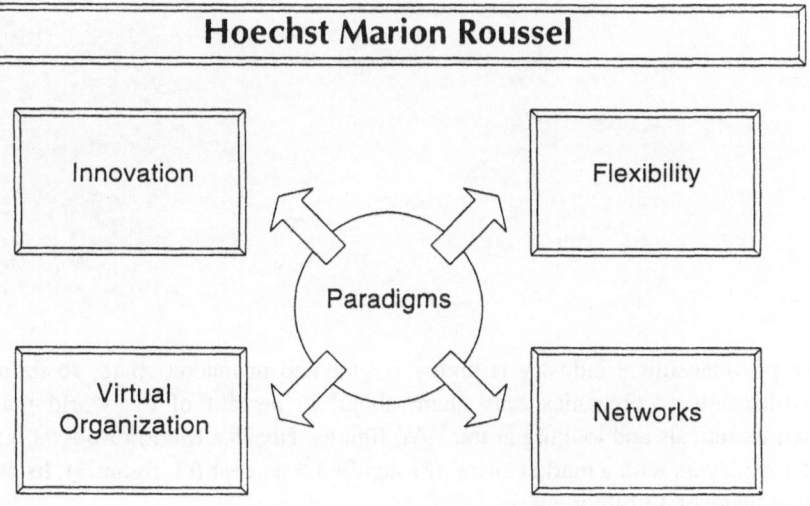

Hoechst Marion Roussel

Because of the very quickly growing knowledge in particular in academia and in the biotech-world it is very critical to operate in a virtual organisation, and to access innovation where it happens. Flexibility is a very important component, flexibility in moving resources, moving people, but also in engaging and disengaging from relationships and partnerships in order to remain at the cutting edge in an innovative research organisation. And lastly, of course, all of that implies networks, efficient networks where a very major component is communication as well as mobility.

Hoechst Marion Roussel is a company that was created in 1995, and is now the pharmaceutical company of the Hoechst group, with headquarters in Frankfurt, Germany. It is actually a composite of what used to be the pharmaceutical division of Hoechst AG in Germany, the pharmaceutical activities of Roussel Uclaf in Paris and the American pharmaceutical company Marion Merril Dow. Figure 2 gives a few highlights of what Hoechst Marion Roussel is all about and what it stands for.

Figure 2:   Global Strategy of Hoechst Marion Roussel

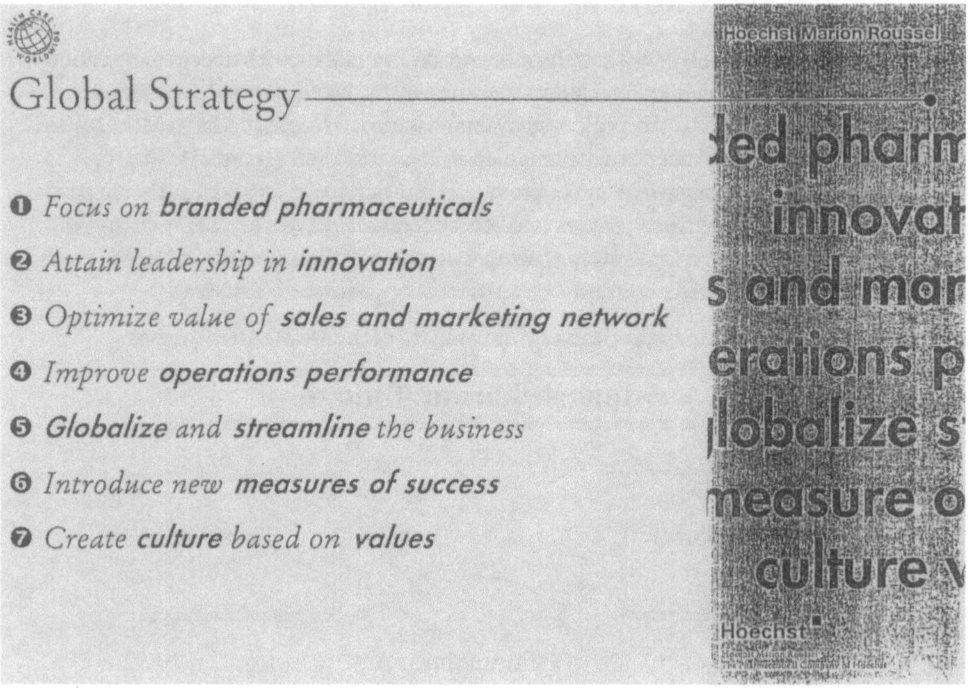

The pharmaceutical industry is highly fragmented in market share, so the top ten pharmaceutical companies only share about 30 percent of the world market in pharmaceuticals and looking at the 1996 figures, Hoechst Marion Roussel is among the top players with a market share of roughly 3.5 percent (cf. figure 3). Its sales are in the range of 13 billion marks.

A very important figure on the same transparency is research and development costs which are enormously high in this industry. Hoechst Marion Roussel devotes in the range of 2.1 billion marks every year towards research and development of new therapies.

Figure 3:   Top Pharmaceutical Companies World-wide
                   by Market Share in 1996

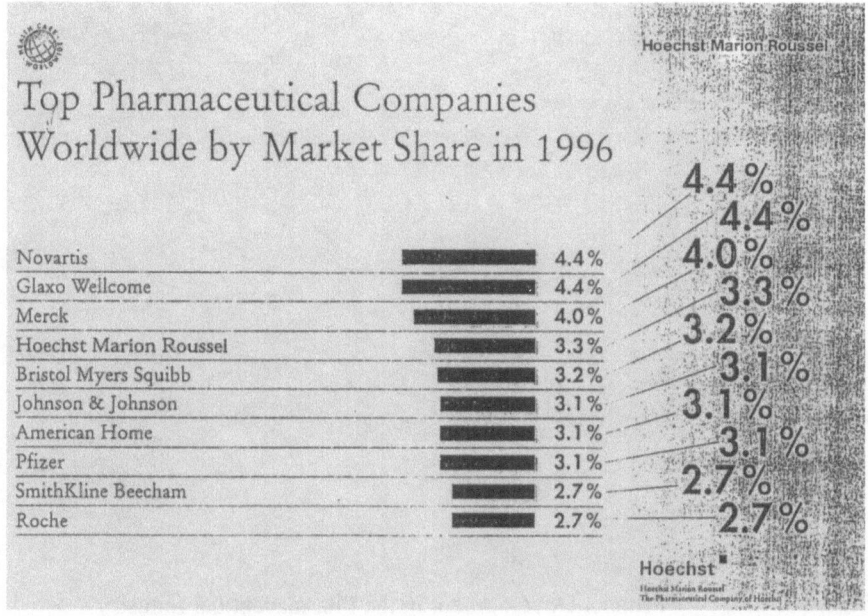

Hoechst Marion Roussel research is an example of a global effort, that is executed locally. This organisation has certain competence centres, and they are Chemistry, Core Research Functions, such as screening or biophysics, Biotechnology, as well as Disease Groups or Therapeutic Areas in which methodologies are applied to find new therapeutics (cf. figure 4).

These centres of expertise are present in several countries, which provide an infrastructure or resources that the company has a particular desire to participate in.

Innovation is the prime driver of many industries and certainly also of the pharmaceutical industry. So at the end of all the research and development, an innovative drug has to be the outcome (cf. figure 5). Otherwise we did not do our business well.

In order to be innovative in drug discovery an enterprise has to understand much better than in the past the cause of diseased, and what the mechanisms are that are

Figure 4:   R&D Locations of Hoechst Marion Roussel

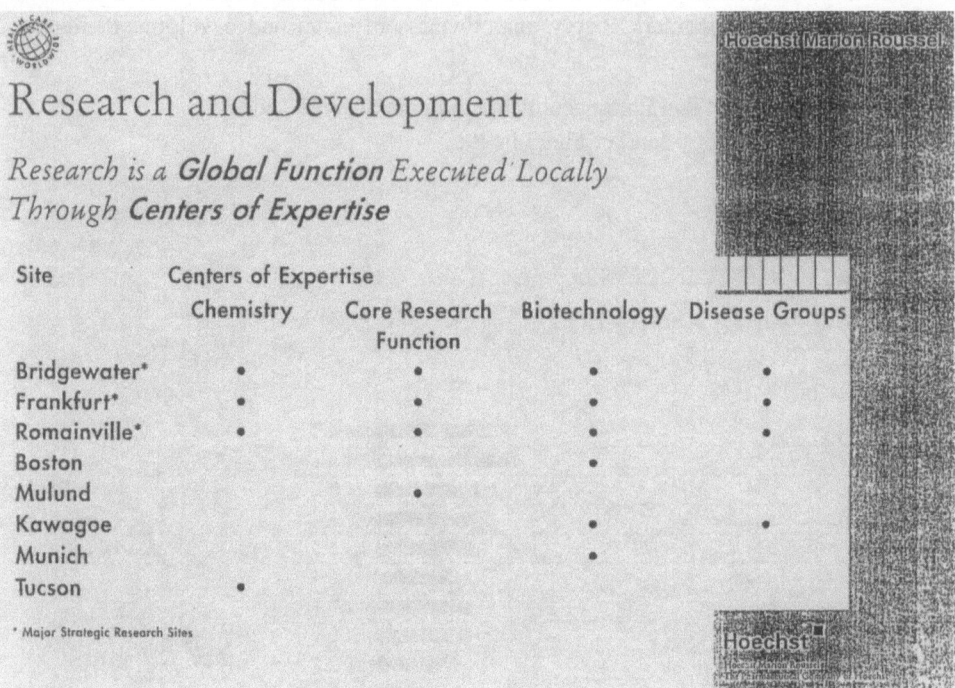

Figure 5:   Product Development in the Pharmaceutical Industry

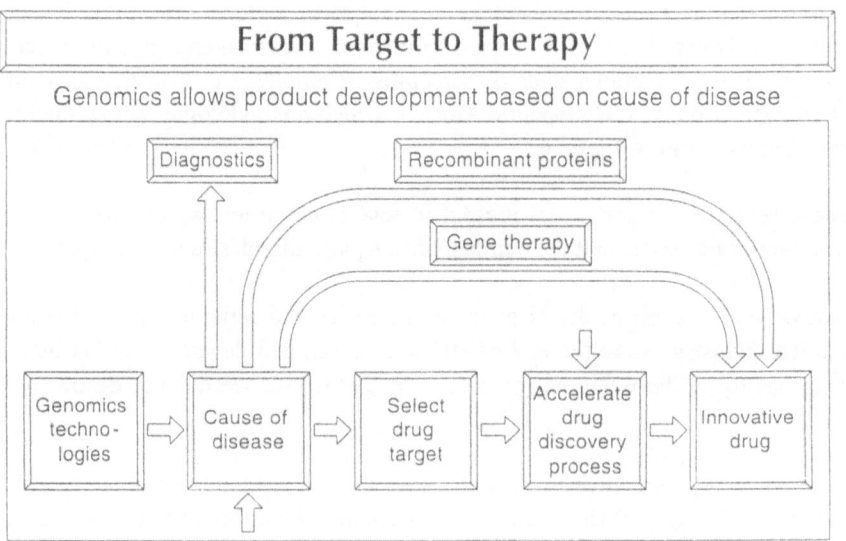

involved in making a tissue a disease tissue and causing illness. Another is that there have been enormous advances, technological advances primarily, in how to accelerate the drug discovery process. A lot of pharmaceutical companies have the same ideas and have the same innovative approaches, and it becomes extremely critical to have the speed to be among the first three on the market in order to truly have a return on investment in pharmaceuticals. So these two areas have significantly changed over the years and where the major paradigm shifts in drug discovery are occurring. The drivers of this industry in the next century will be areas that fall under the umbrella of biotechnology related to understanding disease, and the area of high speed technologies which are more engineering than they are biology. Think of the enormous changes in chemistry that ten years ago would have been unimaginable. Now thousands and thousands of molecules are synthesised within hours and days when previously it took one chemist a year to synthesise 30 compounds. So the changes in this industry are tremendously fast and are driven to a very large extent by a high degree of innovation in technology around the world.

It was mentioned earlier that innovation has to be accessed globally. This is necessary, simply because it is impossible to harness internally all the innovation and all the basic research and all the advances that occur. No organisation is large enough to have sufficient resources and capacity to do that. So Hoechst Marion Roussel has a very strong philosophy of partnering but do that by having internal skills and assets that are attractive so that the companies can indeed be partners of choice for both academic institutions as well as the biotechnology sector (cf. figure 6).

Figure 6:   The Importance of Internal Excellence
                 and Academic and Industry Partnering

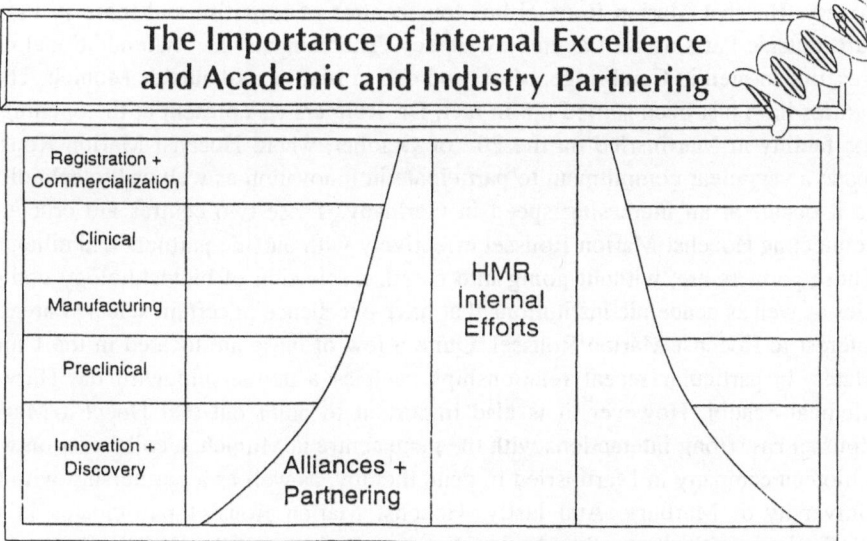

The research part which the enterprise calls innovation and discovery is perhaps the one that relies the most on alliances and partnerships for these reasons.

Figure 7: The Virtual Research Organisation of Hoechst Marion Roussel

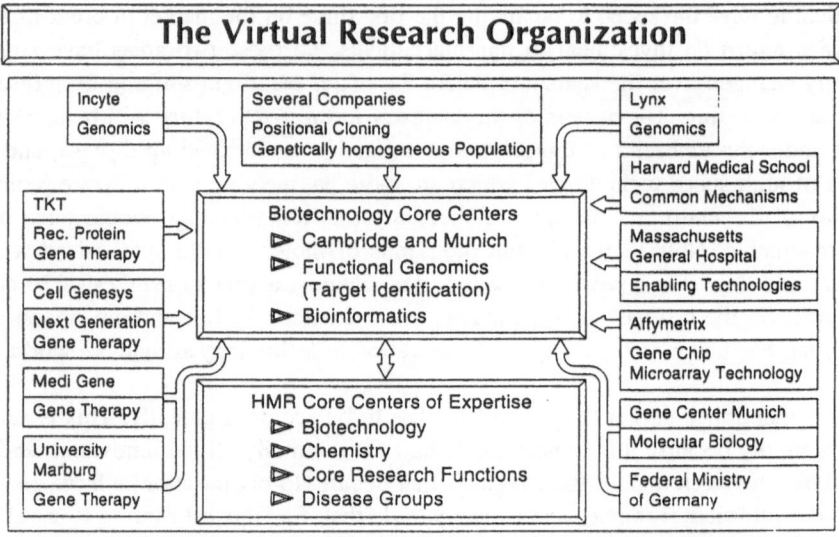

Hoechst Marion Roussel

Figure 7 shows a snapshot of the virtual research organisation currently operating at Hoechst Marion Roussel. It shows some impacts on the economy in Germany of an organisation such as Hoechst Marion Roussel as a major player in research and discovery. Hoechst Marion Roussel has four centres of expertise and those in Frankfurt, outside Paris as well as in New Jersey. The company has two additional centres, one located in Cambridge, Mass. and one in Martinsried outside Munich. These centres have just been started up. In fact, Dr. Rüttgers was present at the opening of the facility in Martinsried on the 20th of October, where Hoechst Marion Roussel made a very clear commitment to participate in innovation as well as biotechnology as it occurs at an increasing speed in Germany. These two centres are crucial in connecting Hoechst Marion Roussel effectively with outside partners and alliances. Those partners are, without going into detail, a selection of biotechnology companies as well as academic institutions that have excellence in certain selected areas of interest to Hoechst Marion Roussel. Quite a few of these are located in the United States, in particular recent relationships such as a partnership with the Harvard Medical School. However, it is also important to point out that Hoechst Marion Roussel has strong interactions with the gene centre in Munich, a collaboration with a biotech company in Martinsried in gene therapy, as well as a partnership with the University of Marburg. And lastly, Hoechst Marion Roussel participates in the BioRegio activities as well as in the "Leading Projects" of the BMBF. So the support provided by the local environment is one that the company takes up very quickly and has already incorporated into its virtual research organisation.

So Germany participates in a technology that will undoubtedly be a revolution in biology and cause a major change in how we look at medical discovery in the future. Biotechnology is a term that is used quite frequently. It describes a collection of enabling technologies that utilise biological information or techniques at the molecular level, so at a very, very small microscopic or sub-microscopical level, towards the discovery and production of innovative and differentiable therapies.

Two examples are considered in the following. One that is already a well established part of biotechnology and has become a very major contributor to a commercial effort and very significant sales in products. This is the production of therapeutic proteins, the oldest part of the biotechnology industry. It started in the early seventies based on academic work, and then in the United States, there was a trend at the end of the seventies and the beginning of the eighties to translate that academic knowledge into commercial opportunities. This idea of translational or applied research is one that is very aggressively followed in the United States. Figure 3 shows a time-line upon which those discoveries made in academia were turned into commercial success by biotechnology companies. As a result, in 1995 alone that part of the industry contributed about eleven billion dollars in sales.

Figure 8:   The Evolution of Technology into Products:
              the Example of Genomics

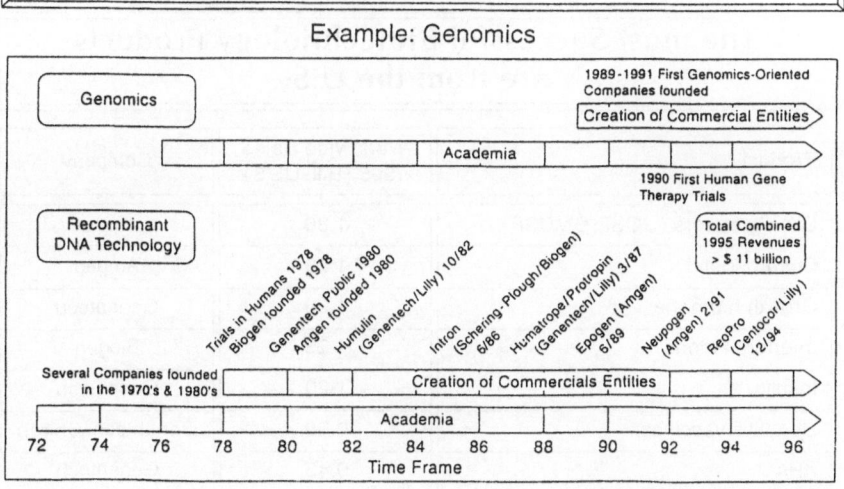

Hoechst Marion Roussel

There is a new area or a new "revolution" in biotechnology that is on-going, called genomics (cf. figure 8). It simply means that scientists are about to determine the entire genetic make-up of every human cell, of lower organisms, of pathogenic organisms, from which a profound knowledge about disease and ways of interfering with disease will be derived. Genomics also has a long history in academia, but its commercial translation has occurred much later, only in the late eighties and in the early nineties, and perhaps most prominent way with what is generally known as gene-therapy that has made a lot of waves and has caused a lot of discussions. But this new way of biotechnology will make a commercial contribution. It will by far exceed the eleven billion dollars that are part of the more established traditional biotechnology arena.

Figure 9 gives a view of successful Biotechnology products and their annual sales in dollars. These are very big sales figures. Without any exception all of those compounds were brought to the market and were brought to commercial success by American-based biotechnology companies. The new wave is no longer limited to recombinant proteins, but research and discovery that is very much coined by the on-going revolution of genomics. And in the concept of virtual organisations and global access there is a very large number of relationships that were formed between large pharma and biotechnology companies in order to exploit the information that is coming out of the so-called genomics revolution. And without going into detail there is no question that these are very substantial relationships (cf. figure 10). To just give a number, in 1996 one billion dollars were invested in genomics, just in the hope and anticipation that this will be the new wave of medical break-through.

Figure 9:   Successful Biotechnology Products

The most Successful Biotechnology Products are from the U.S.

Product	Worldwide Sales 1995 (Bill. US $)	Company
Growth factors (GCSF, GMCSF)	1.80	Amgen
Erythropoetin	1.79	Amgen
Growth hormone	1.28	Genentech
Interferon alpha	1.23	Biogen
Insulin	1.20	Genentech
Hepatitis vaccines	0.80	Genentech/Chiron
rtPA	0.43	Genentech
Interferon beta	0.42	Biogen

Source: Wood, MacKenzie 1995

Hoechst Marion Roussel

Figure 10:   A Genomic View of the World

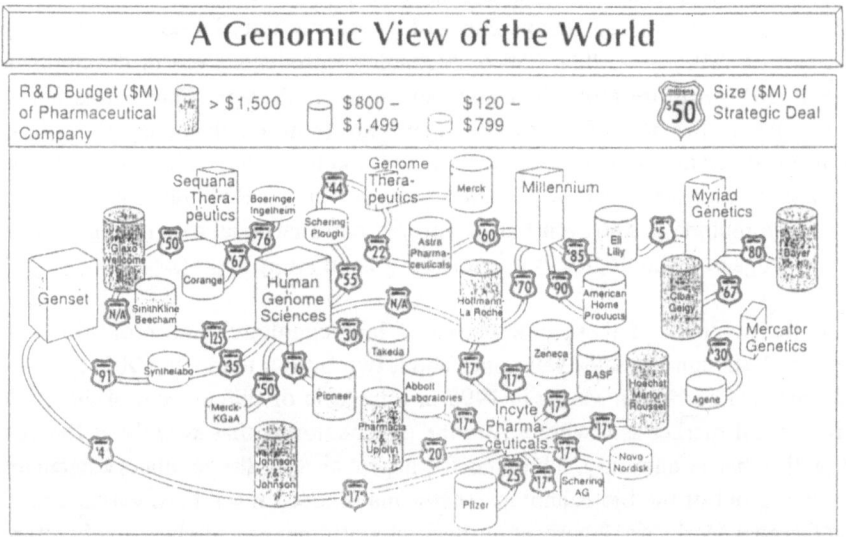

Hoechst Marion Roussel

Figure 11:   Biotechnology Drugs in Development

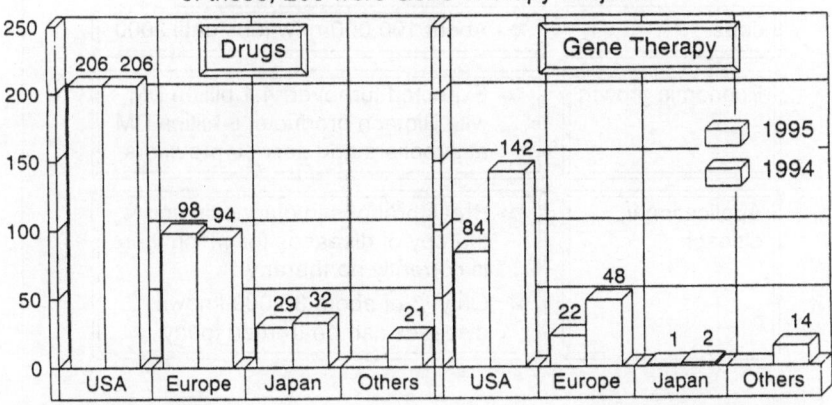

These figures show that biotechnology no longer just means therapeutic proteins, it is small molecules, it is methodologies, it is screening, it is combinatorial chemistry, it is the whole spectrum, it is the paradigm of the new century in medical discovery. At the end of 1995, more than 750 biotechnology products were in development (cf. figure 11). Of those more than 200 were in gene therapy. That leaves a very large fraction in the area of approaches that were traditionally done by the large pharma concerns and are now being done in addition to the large pharma also by the biotechnology industry. And the United States are leading that industry by far. All of Europe combined is not even half as strong, and that is as true for the smaller molecule drugs or technologies as it is true for gene therapy approaches.

What is the potential of these new waves of biotechnology? The projections are very clear. Although the sales were only about two billion marks in 1985, they had already reached 15 billion Marks in 1995 or a fraction of five percent of the world pharmaceutical market as a whole. And the expectation is that as early as the year 2000, with what is currently submitted for approval with the regulatory agencies, sales coming out of the biotechnology sector may indeed reach between 85 and 90 billion German Marks or 17 percent of the total world pharma market. So, therefore, Germany cannot afford not to play in the biotechnology arena and not to be part of this industry. The reasons are very clear (cf. figure 12). First of all, there are tremendous job opportunities. It will certainly be a substantial number.

Figure 12:   Gene- and Biotechnology is a Technology of the Future

Gene- and Biotechnology is Technology of the Future

Three reasons, why Germany must not do without it

Jobs	▷ About 100,000 new jobs until 2000
Economic growth	▷ Expected turnover: 4.1 billion DM with Biotech products, 8 billion DM at suppliers and service providers
Application in disease	▷ Breakthroughs in diagnosis and therapy of diseases for which there is currently no therapy ▷ Only 1/3 of about 30,000 known diseases can be treated today

Hoechst Marion Roussel

There is also enormous economic potential in an expected turnover of about four billion Marks with biotech products and another eight billion Marks simply through providers and suppliers of services to that particular industry. And lastly, there is an obligation to participate because there are many areas of therapeutic need for which one currently has no treatment. And breakthroughs are desperately required because only about one third of the 30,000 known diseases can currently be treated, whereas the rest cannot and we are still looking for adequate therapeutic intervention.

Figure 13:   A Selection of Positive Signals for Gene Technology in Germany

Gene Technology – Germany on the Rise
A Selection of Positive Signals

Venture capital is being mobilized

▷ BioRegio - competition 500 Million DM
   Venture funds: Bayrische Vereinsbank, ING-Bankengruppe (NL), Bayer,
                  Boehringer Mannheim 150 Million DM

Investments of the Pharmaceutical industry in basic research and cooperations are increasing

▷ HMR Core Research Center with focus on genomic research in Martinsried

Number of Biotech companies has doubled from 75 (1995) to 150 (1996)

Strategic alliances between German and American companies are increasing

▷ Morphosys (Munich) – Pharmacia Upjohn; Evotec (Hamburg) – Novartis, SmithKline Beecham

New Biotech products are produced in Germany

▷ Reteplase (Boehringer Mannheim), Saruplase (Grünenthal), Lepirudin (HMR)

QUIAGEN is the first company listed on NASDAQ
(US stock exchange for High-Tech-Companies)

Hoechst Marion Roussel

There are a lot of positive signals in Germany (cf. figure 13). Some of them were already touched on in one way or another. One is that venture capital is finally being mobilised. BioRegio, as was mentioned this morning by the federal minister, has given this whole effort a major boost by making available about 500 million marks to this effort. There are also many venture funds that are finally coming in, there are investments that are being made by the pharmaceutical industry like Hoechst Marion Roussel. Already mentioned was the biotechnology centre the company just started in Martinsried in October. It was said earlier by the research minister that the number of biotech companies has doubled from about 75 to 150 in one year and the trend is increasing. And alliances are being formed between US-based or international pharma companies and German biotech companies.

Products are being produced in Germany and although it is only one company so far, the biotechnology industry is becoming successful enough to be listed on the New York exchanges, the Nasdaq in this case.

There is a stated goal that Germany wants to achieve the number one position in biotechnology in Europe by the year 2000 (cf. figure 14).

Figure 14:   Biotechnology in Germany:
                   Goals, Conditions and Prerequisites

Biotechnology in Germany
Goal: Number 1 in Europe by 2000

Conditions and prerequisites

▷ Scientific potential and research infrastructure

▷ Cooperation of science and industry

▷ Venture capital for financing of start-ups

▷ Effective protection of intellectual property

▷ Efficient approval of applications to operate facilities

▷ Appropriate pricing and reimbursement for innovation

▷ Acceptance in the population

Hoechst Marion Roussel

This is a very ambitious goal. To achieve it, the proper environment and infrastructure are necessary plus a few other things which are particularly important. One is an enormous scientific potential. In Germany there is the appropriate research infrastructure. These are very important prerequisites. Co-operation between academia and industry has to be improved and money is needed to start up companies, encourage entrepreneurship and facilitate the process of translating research into application. Universities have to improve themselves in particular the protection of intellectual property, because that is after all the major asset and lifeblood of a biotech company. An environment is needed in which it is relatively easy to get approval to operate facilities in the biotech arena, and, of course, pricing and reimbursement for innovation are very important parameters. Nowadays the acceptance rate in Germany is as high as is needed in order to implement Biotechnology broadly and to participate in this very important scientific and business opportunity.

Policies to Strengthen Innovative Start-ups and SMEs in Global Competition

Falk Strascheg

One can divide companies into three categories:

- mice
- gazelles
- elephants

Mice are companies which will always stay small and never grow. Gazelles companies venture capitalists are looking for: These companies start small but ha the potential to grow <u>and</u> use their potential. Elephants are large established co panies which are not as flexible as small ones. This paper would like to present h to create gazelles in Germany or in Europe.

Figure 1:   Comparison of Young and Old Companies
              with Sales above $ 50 Million

Comparison of young and old companies with sales above $ 50 million

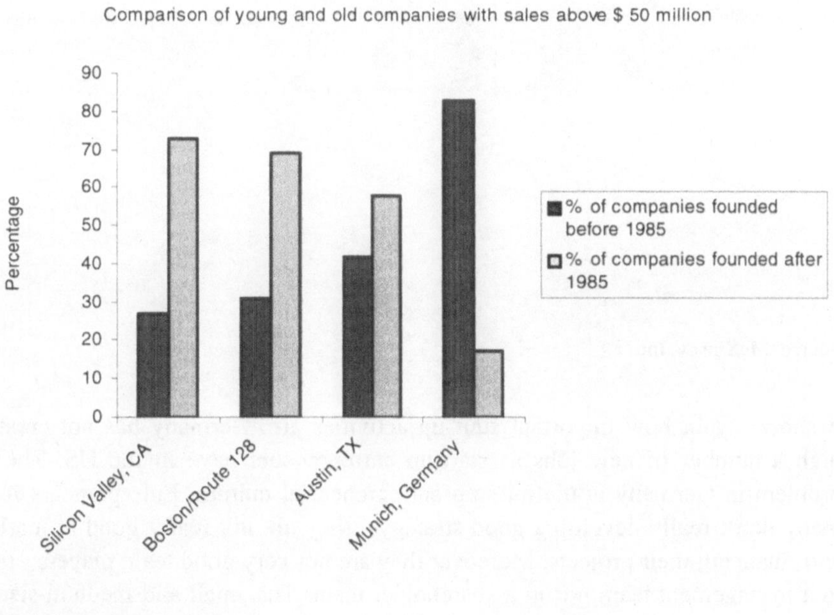

Source: McKinsey & Company, Inc.

Figure 1 shows companies with more than 50 million dollar sales in different regions. Looking at the Silicon Valley, only 27 per cent of these companies were
founded before 1985. In Munich, the "Silicon Valley" of Germany and one of the
most advanced technology centres in this country, 83 per cent of the companies
with sales of more than 50 million dollars were founded before 1985. So there is a
lot of ground to make up, compared to the American competitors.

In figure 2, which was compiled by McKinsey, it can also be realised that in the
highly technological Boston area about 80 per cent of all new jobs are created by
start-up activities, and only about 10 to 20 per cent are expansions of existing companies and very few of them are investments by outside companies.

Figure 2:   Impact Start-up Sctivities / Net Job Creation 1992 - 1996

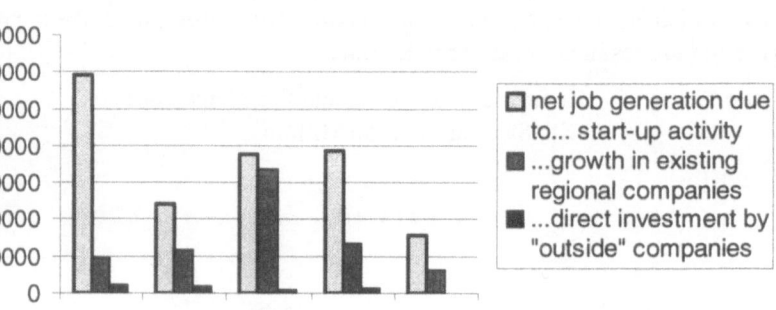

Start-up activities have the far hightest impact on net job creation -
examples of regions 1992-96

Source: McKinsey, Inc.

It shows again how important start-up activities are. Germany has not created as
high a number of new jobs as start-up entrepreneurs have in the US. The main
problem in Germany is the missing entrepreneurial culture. Entrepreneurs in Germany don't really develop a good strategy - they are not really good at marketing
and financing their projects. Moreover they are not very good team players - neither
in a management team nor in a shareholder team. The small and medium-sized enterprises in Germany have only one per cent of external shareholders. These are
shareholders who neither work in the company nor are related to people working in
the company. In Germany external shareholders are almost considered as

"invaders". This is different to cultures in other European countries: In Europe an average of about 20 per cent of all shareholders in small and medium-sized enterprises are external shareholders. In the Netherlands even 36 per cent of all shareholders are external shareholders.

There are some significant differences between German and American entrepreneurs. American entrepreneurs found companies to make money while German entrepreneurs found businesses for self-fulfilment. Most people in Germany just didn't want to have a boss anymore.

The Americans really aim to become global players and make strategic plans for considerable growth. In Germany people found businesses the size of - what we call – "Ingenieurbüro" type, i.e. very small enterprises. In the USA a company is primarily founded with sufficient equity, including all the costs which are needed to become a global player. In Germany a lot of people start with very small amounts of equity.

In the USA people are ready to make cash when they have grown up their company. Here we have a "Blut-Erde-Scholle"-mentality, which is very hard to translate into English. It means people are bound to their soil – they think in generations. The next manager of the company will have to be the son and the next a grandson. This is not too healthy for business.

One more difference is the fact that Americans enjoy their wealth and they are proud of it – in Germany wealth is being hidden.

What are the reasons for the Germans' weaknesses? Certainly the educational system, but it depends also a lot on society and its values. Moreover the welfare system is very anti-risk and, therefore, people are not prepared to take risks. Sometimes there are too few visions while lack of money is the reason for missing high growth companies.

One very good example of good marketing of the Americans is MIT. The Boston Bank made a study on behalf of MIT analysing how many jobs were created by graduates of MIT since 1900:

Figure 3 shows that graduates of MIT have created 1.1 million jobs in founding about 4,000 companies with sales of about 230 billion US dollars. Looking at the last years the figures are very impressive. However, if you analyse the study carefully, you find out the following: 60% of the companies were founded after 1980 – but the companies which are really big from employees' point of view (Campbells Soup or similar companies) had already been founded in 1900 or 1920. Even if figures for the last year look very impressive these American marketing efforts should be looked at cautiously.

Figure 3:   MIT Spin-off Companies

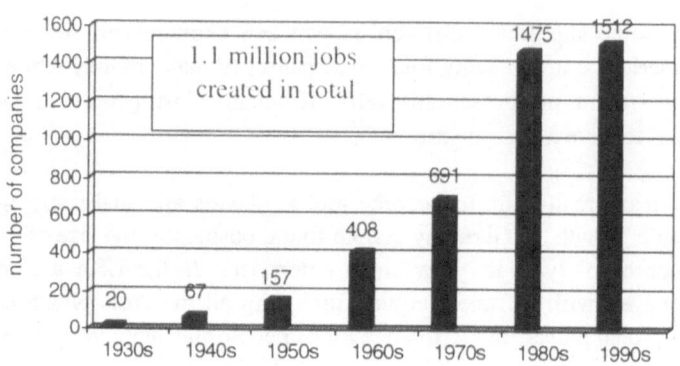

Source: MIT, "The impact of innovation", 1997

What does the venture capital situation look like in Europe? In 1996 the European
venture capital companies invested about DM 13 billion in companies located in
different European countries probably - this figure will be much higher in 1997
(figure 4).

Figure 4:   European Venture Capital Investments 1988 - 1996

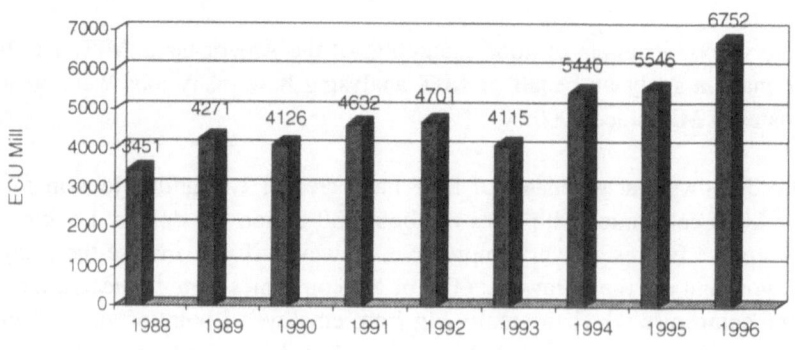

Source: EVCA

Which role is Germany playing in this game? Germany still plays an unimportant part – in 1996 just about DM 1.37 billion were invested.

Figure 5:   Amount Invested 1996 by Country

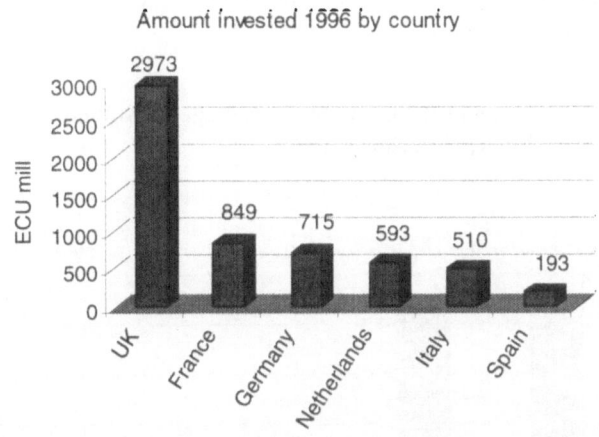

Source: EVCA

But the distribution of this amount is quite favourable towards young high-tech companies.

Figure 6:   Amount Invested per Capita (ECU)

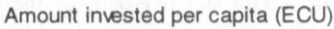
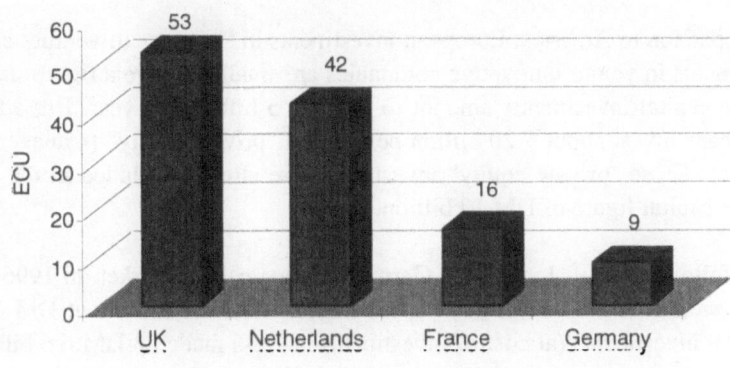

Source: Technologieholding VC GmbH

Figure 6 shows the amount invested per capita in different countries and compares figures from the UK, France, the Netherlands and Germany.

Venture capital investments in Germany make up only ECU 9 per capita (ca. DM 18 per capita) – venture capital investments in the UK are almost six times higher, in the Netherlands they are almost five times higher and in France they are nearly twice as much as in Germany.

Figure 7:   Stage Distribution by Percentage of Amount
                   Invested in Europe 1996

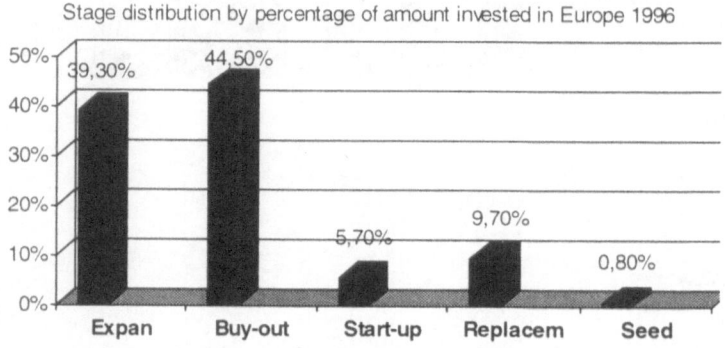

Source: EVCA

Figure 7 shows that in Europe investments in young companies (start-up and seed investments) only make up about 6.5 per cent of the total venture capital investment.

In comparison to America, European investments in the sense of venture capital i. e. investments in young innovative companies are small. To give a figure: In America venture capital investments amount to about $ 5 billion per year. But additionally Americans invest about $ 20 billion per year in "private equity" (management buy-out, etc). Those "private equity" investments are already included in the European venture capital figure of DM 13 billion.

In the following let us look at the German venture capital market. In 1996 DM 1.37 billion were invested in venture capital. Funds available are about DM 10 billion and total investments (at cost of investment) in 1996 made up DM 6.6 billion. How did the money get invested? The area which is interesting is the start-up and seed area which is about 10 per cent of the total investments. Within the next years this percentage will increase. According to the semi-annual report of the German venture capital association, investments in the first half of 1997 more than doubled compared to the first half of 1996.

In Germany for early stage investments very good schemes are provided by several public bodies. Especially from the BMBF (German Ministry for Innovation and Technology), which set up the BTU program. This program supports participation schemes in innovative technology companies. There are two different ways the program is being implemented: On one hand there is the "Deutsche Ausgleichsbank" (tbg) offering a co-investment model with venture capitalists. On the other hand there is the "Kreditanstalt für Wiederaufbau" (KfW) offering a refinancing scheme. The KfW also provides a "risk capital program". For investments in the "new German states" there is a special fund supplied by the KfW. In some of the federal states there are local schemes, for example "Bayernkapital" in Bavaria. Due to these schemes venture capital companies have a good leverage.

How can the situation be improved? It is very important:

- to teach entrepreneurship at universities
- to enforce private venture capital investors
- to create the right climate for entrepreneurs.

What can universities do to teach people how to become entrepreneurs? They can:

- give them a better education on financing
- give them more information on how to set up a company
- teach them how to run a company - setting-up is only a short period of time, but afterwards you have to run it.

Case studies could be done or people could be sent to other companies for traineeships before they start to set up their own company.

It is also important to support private venture capital investors. This can be done by:

- leveraging venture capital investments as done by the BTU program. Parallel loans, guarantees and SBIC schemes as used in America could be very favourable
- promoting venture capital management companies
- restricting competitive government programs. Unfortunately the German federal states have started their own programs which compete with private venture capital. This could really kill private initiatives in this sector
- improving the tax situation for funds in Germany

At the moment funds are fully taxed for all capital gains in Germany. And, of course, a foreign investor is not interested to pay 50 to 60 per cent or more of the returns. Transparent funds should be created which means that funds themselves are not subject to tax but the tax payable would be charged on the shareholder level. To give an idea: 70 to 80 per cent of the DM 450 million which are under management of my company come from foreign investors. Taxes are an important question. So at the moment, funds of the company are incorporated in the Netherlands, which creates some high-paid jobs in the Netherlands, but not in Germany. I would be pleased if we did not have to found Dutch companies.

It is also important to have the right climate for entrepreneurs and to create a new spirit as was done in the MIT business plan competitions, which really generated an entrepreneurial culture. There was a similar program initiated by McKinsey which started in mid 1996 and ended in mid 1997. This pilot program was extremely successful. There were about 150 applicants each in the two cities where it was carried through (Berlin and Munich). In Munich, for example, there will be 20 to 30 good company foundations financed by venture capital. There are following initiatives in 1997 and 1998 in Germany and in some other European countries, for example in the UK (London, Cambridge), Austria (Vienna), Switzerland (Zurich), France (Paris).

The entrepreneurial process is based on four elements:

- the idea
- the entrepreneur and the management team
- the network community
- venture capital.

If these elements exist and really co-operate it is most probable that the company will be successful.

If all things which are outlined in this text are followed, it should not be too difficult that e.g. the (technical) university of Munich or any other research institute could at some time show a similar picture like MIT. It could be quite important to have some vital organisations which also enforce progress in this matter in universities to make some very dynamic gazelles out of them.

The Authors

Professor Ulrich Hiemenz, Director, OECD Development Centre

Dr. Hans-Georg Junginger, Vice President, R&D, Sony Europe GmbH

Dr. Andre Jungmittag, Senior Researcher, Fraunhofer Institute for Systems and Innovation Research (ISI)

Professor Frieder Meyer-Krahmer, Director, Fraunhofer Institute for Systems and Innovation Research (ISI); Université Louis Pasteur, Strasbourg

Professor Jürgen Mittelstrass, Universität Konstanz; Council for Research, Technology and Innovation

Professor Yrjö Neuvo, Senior Vice President, R&D, Nokia Mobile Phones

Dr. Sylvia Ostry, Distinguished Research Fellow, Centre for International Studies, University of Toronto

Dr. Guido Reger, Senior Researcher, Fraunhofer Institute for Systems and Innovation Research (ISI)

Norbert G. Riedel Ph. D., Vice President, Global Core Research Functions and Global Biotechnology, Hoechst Marion Roussel Inc.

Dr. Jürgen Rüttgers, Federal Minister, German Ministry of Education, Science, Research and Technology

Professor Luc Soete, Director, Maastricht Economic Research Institute on Innovation and Technology (MERIT)

Falk Strascheg, Technologieholding VC GmbH, Chairman of the European Venture Capital Association (EVCA)

TECHNOLOGY, INNOVATION and POLICY

Series of the Fraunhofer Institute
for Systems and Innovation Research (ISI)

Springer
and the
environment

At Springer we firmly believe that an international science publisher has a special obligation to the environment, and our corporate policies consistently reflect this conviction.
We also expect our business partners – paper mills, printers, packaging manufacturers, etc. – to commit themselves to using materials and production processes that do not harm the environment. The paper in this book is made from low- or no-chlorine pulp and is acid free, in conformance with international standards for paper permanency.

 Springer

Druck: betz-druck GmbH, D-64291 Darmstadt
Verarbeitung: Buchbinderei Schäffer, D-67269 Grünstadt